SpringerBriefs in Environmental Science

More information about this series at http://www.springer.com/series/8868

Genesis T. Yengoh • David Dent
Lennart Olsson • Anna E. Tengberg
Compton J. Tucker III

Use of the Normalized Difference Vegetation Index (NDVI) to Assess Land Degradation at Multiple Scales

Current Status, Future Trends, and Practical Considerations

Springer

Genesis T. Yengoh
Lund University Centre for Sustainability
    Studies - LUCSUS
Lund, Sweden

David Dent
Chestnut Tree Farm, Forncett End
Northfolk, UK

Lennart Olsson
Lund University Centre for Sustainability
    Studies - LUCSUS
Lund, Sweden

Anna E. Tengberg
Lund University Centre for Sustainability
    Studies - LUCSUS
Lund, Sweden

Compton J. Tucker III
Department of Hydrospheric
    and Biospheric Sciences
NASA Goddard Space Flight Center
Greenbelt, MA, USA

ISSN 2191-5547   ISSN 2191-5555   (electronic)
SpringerBriefs in Environmental Science
ISBN 978-3-319-24110-4   ISBN 978-3-319-24112-8   (eBook)
DOI 10.1007/978-3-319-24112-8

Library of Congress Control Number: 2015952316

Springer Cham Heidelberg New York Dordrecht London

Springer International Publishing AG Switzerland is part of Springer Science+Business Media (www.springer.com)

Preface

This report examines the scientific basis for the use of remotely sensed data, particularly the normalized difference vegetation index (NDVI), for the assessment of land degradation at different scales and for a range of applications, including resilience of agroecosystems. Evidence is drawn from a wide range of investigations, primarily from the scientific peer-reviewed literature but also non-journal sources. The literature review has been corroborated by interviews with leading specialists in the field (Chap. 7).

The use of continuous time series of global NDVI data, based on the NOAA AVHRR sensor, developed rapidly in the early 1990s. Since then, data processing and techniques for analyses of the data have improved significantly. Developments in data quality screening, geometric correction, calibration between sensors, atmospheric and solar zenith corrections, cloud screening, and data mosaicking have enabled the production of several databases of global NDVI data of high quality that are freely accessible over the Internet. The spatial resolution of these datasets ranges from coarse (8–1 km) to medium (250 m).

Even if there is no alternative to remotely sensed data for global- and continental-scale monitoring of vegetation dynamics, the technique is not without weaknesses. The report reviews the use of NDVI for a range of themes related to land degradation. Land-cover change, including deforestation, has been detected quantitatively even though the drivers are elusive (Sect. 3.1). Drought monitoring and early warning systems use NDVI data and have developed fully operational systems for data dissemination and analysis (Sect. 3.2). Desertification processes at the global, continental, and subcontinental scale have been studied intensively in the last two decades; a key finding is that most of the world's drylands show a trend of increasing NDVI. Interpretation of the causes and implications of that greening trend is still a matter of discussion (Sect. 3.3). Soil erosion has been studied at national and sub-national level, primarily using NDVI derived from data of medium to high spatial resolution, such as MODIS at 250 m and Landsat at 30 m resolution (Sect. 3.4). Salinization of soils has been studied using a wide range of remotely sensed data; however, its detection and monitoring is more experimental than for other forms of land degradation (Sect. 3.5). Monitoring of vegetation burning is a field where NDVI has been widely used for three purposes: to assess the risk of fire in terms of fire-fuel load, detection and monitoring of fire, and mapping burned areas, as well as monitoring vegetation recovery after fire (Sect. 3.6). Soil carbon is an important indicator of land productivity that has been studied by means of NDVI (Sect. 3.7), both as a stand-alone measure and as input to ecosystem models. Studies have shown a high agreement between NDVI-based and ground-based estimates. NDVI has been used to facilitate mapping and monitoring of biodiversity loss, even though species composition cannot be assessed (Sect. 3.8).

Part of the Terms of Reference of the report was to assess the potential of NDVI for monitoring of agroecosystem resilience. Resilience is a concept with a broad range of definitions. In its original form, the ability to recover from disturbance or stress, ecosystem resilience can be assessed to a certain degree by combining NDVI with ancillary data, such as rainfall, often described as a hysteresis curve (Sect. 3.9).

Even if NDVI is by far the most commonly used vegetation index, other indices have been proposed and used for global-scale studies, such as two types of the Enhanced Vegetation Index (EVI). These indices are reviewed and compared with NDVI. The 3-band EVI is subject to certain technical problems, and the 2-band EVI is highly correlated with NDVI. Our conclusion that NDVI is the preferred index for operational monitoring is thus strengthened by the comparison (Sect. 2.3).

Interpretation of trends and patterns of NDVI data cannot automatically be interpreted in terms of land degradation and improvement. The report highlights a number of issues that must be considered for operational use of NDVI for land degradation monitoring (Chap. 4).

It is sometimes asserted that NDVI becomes saturated for dense vegetation (leaf area index >1). This is incorrect. The reason for the asymptotic relationship between NDVI and LAI is because all visible photons are absorbed at high LAI. In fact this confirms that NDVI represents photosynthetic capacity and primary production rather than LAI (Chap. 4).

Distinction between land degradation/improvement and the effects of climate variation is an important and contentious issue. There is no simple and straightforward way to disentangle these two effects. Rain-use efficiency (RUE), calculated by dividing NDVI by rainfall, is sometimes used to separate human action from natural variation. Even if theoretically sound, there are both technical and scientific problems with this approach. The technical problems are related to the mismatch of scale between climate data (most often point based) and NDVI (spatially continuous). Spatial interpolation of point observations is highly problematic, at least for short time periods. The scientific problems are concerned with the contextual relationship between vegetation and rainfall. The following rule of thumb may be applied: where vegetation dynamics are strongly driven by rainfall, i.e., in drylands, declining RUE is correlated with land degradation. In humid areas, where vegetation is not as strongly driven by variations in rainfall, NDVI in itself is strongly correlated with vegetation dynamics and may be taken as a proxy for land degradation and improvement provided that potential false alarms are accounted for (Chap. 4, Sect. 5.2).

The accessibility and reliability of datasets is crucial for operational monitoring. The report reviews 14 of the most important datasets of NDVI and six climate databases that potentially can be used in combination with NDVI data. The most widely used, and also the most rigorously tested dataset is the GIMMS (Global Inventory for Mapping and Modeling Studies) and its most recent version the GIMMS3g. It comprises continuous data coverage from August 1981 to the present at 15-day intervals. It is the only dataset that is continuously updated with new data. The GIMMS3g is accessible free of charge over the Internet. Since 2000, NDVI data are available from both AVHRR (at 8-km resolution) and MODIS (1000–250-m resolution). Normally they are highly correlated, but we recommend the use of MODIS data as the benchmark (Chap. 8).

For monitoring at national, subnational, and project levels, the report recommends the use of nested approaches in which coarse-resolution data, such as AVHRR NDVI at 8-km resolution, are combined with other remotely sensed data that offer higher spatial resolution ranging from 0.5 to 250 m and better spectral and radiometric resolution. These data, calibrated against the long-term but coarse-resolution AHVRR database, can be used to elucidate reasons for changes in the coarse-resolution NDVI signal (such as forest destruction or other land-use change, habitat fragmentation, or soil erosion) and for national, subnational reporting, and project monitoring (Chap. 9).

A combined approach—analyzing spatial patterns and temporal trends in the coarse-resolution imagery, zooming in for greater detail using fine-resolution NDVI supplemented by systematic information on climate, terrain and land cover, and spot checks with very high-resolution commercial satellite imagery—can contribute to the assessment of agroecosystem resilience. Targeted research is needed to establish exact procedures for such applications (Chap. 9).

Although NDVI data are easily accessible and free of charge, successful monitoring of land degradation requires adequate technical, institutional, and skilled human resources. Such capacity, however, can probably be built effectively at existing regional and/or national centers (Chap. 10). With such capacity, NDVI can be used for cost-effective and reliable national reporting on several of the UNCCD core indicators (Sect. 11.1) and potentially also as input to a revised GEF resource allocation method, at least after some further testing on real data (Sect. 11.2).

To conclude, a substantial body of peer-reviewed research lends unequivocal support for the use of coarse-resolution time series of NDVI data for studying vegetation dynamics at global, continental, and subcontinental levels. There is compelling evidence that these data are highly correlated with biophysically meaningful vegetation characteristics such as photosynthetic capacity and primary production that are closely related to land degradation and to agroecosystem resilience. The

GIMMS3g dataset that now contains continuous data coverage since August 1981 is the most reliable, used, and cited database as well as the only database that is up to date, free of charge, and will be continued for the foreseeable future.

Lund, Sweden	Genesis T. Yengoh
Northfolk, UK	David Dent
Lund, Sweden	Lennart Olsson
Lund, Sweden	Anna E. Tengberg
Greenbelt, MA, USA	Compton J. Tucker III

Acknowledgments

This report benefited from the inputs of Guadalupe Duron (STAP Secretariat), Annette Cowie (STAP member for land degradation), Victor Castillo (UNCCD Secretariat), Mohamed Bakarr (GEF Secretariat), Thomas Hammond (STAP Secretariat), and Virginia Gorsevski (STAP Secretariat). We are also very grateful for the scientific and technical advice, as well as support with valuable documentation, from Stephen Prince (University of Maryland), Stefanie Herrmann (University of Arizona), Alfredo Huete (University of Technology Sydney), Graciela Metternicht (University of New South Wales), Alexander van Oudenhoven (Wageningen University), Maria Luisa Paracchini (European Commission, Joint Research Centre), Anupam Anand (University of Maryland), Rasmus Fensholt (University of Copenhagen), Barron Joseph Orr (University of Alicante), Jeroen E. Huising (International Institute of Tropical Agriculture, IITA), and Keith Shepherd (Columbia University). Material that makes up the annexes of this report and contributes substantially to different sections was developed by the following scientist, to whom we are indebted: Michael Cherlet, Hrvoje Kutnjak, Marek Smid, and Stefan Sommer (European Commission, Joint Research Centre); Eva Ivitis (European Environment Agency); Zhanguo Bai (World Soil Information-ISRIC, Wageningen); and Kebin Zhang (AO-LADA Task Force and TPN3 Focal Point of UNCCD, China). We also thank GISGeography.com for their contribution with a summary of commonly used open-source geospatial software.

Report commissioned by the Scientific and Technical Advisory Panel of the Global Environment Facility (STAP/GEF).

Contents

List of Figures

Acronyms

ASTER	Advanced Spaceborne Thermal Emission and Reflection Radiometer
AVHRR	Advanced Very High-Resolution Radiometer
CIESIN	Center for International Earth Science Information Network
CRU	Climatic Research Unit, University of East Anglia
EUE	Energy-use efficiency
EVI	Enhanced Vegetation Index
FAO	Food and Agriculture Organization of the United Nations
FASIR	Fourier-Adjusted, Sensor and Solar zenith angle corrected, Interpolated, Reconstructed
GDP	Gross domestic product
GEF	Global Environment Facility
GIMMS	Global Inventory for Mapping and Modeling Studies
GLADA	Global assessment of land degradation and improvement
GOME-2	Global Ozone Monitoring Experiment-2
GOSAT	Greenhouse gases Observing SATellite
HANTS	Harmonic ANalysis of Time Series
IRS	Indian Remote Sensing
LADA	Land Degradation Assessment in Drylands
LAI	Leaf area index
LTD	Long-Term Data Record
LUE	Light-Use Efficiency
LULCC	Land-use and land-cover change
MERIS	MEdium-Resolution Imaging Spectrometer
MODIS	Moderate-Resolution Imaging Spectroradiometer
MVC	Maximum Value Composition
NASA	National Aeronautics and Space Administration
NDVI	Normalized difference vegetation index
NDWI	Normalized difference water index
NOAA	National Oceanic and Atmospheric Administration
NPP	Net primary productivity
PAL	Pathfinder AVHRR land

REDD	Reducing emissions from deforestation and forest degradation
RESTREND	Residual trend
RUE	Rain-use efficiency
RUSLE	Revised Universal Soil Loss Equation
SCIAMACHY	SCanning Imaging Absorption spectroMeter for Atmospheric CHartographY
SLM	Sustainable land management
SOC	Soil organic carbon
SOTER	Soil and Terrain Database
SPOT–VGT	Satellite Pour l'Observation de la Terre — Végétation
STAP	Scientific and Technical Advisory Panel of the Global Environment Facility
STAR	System for Transparent Allocation of Resources
UNCCD	United Nations Convention to Combat Desertification
UNEP	United Nations Environmental Programme
UNFCCC	United Nations Framework Convention on Climate Change
VASClimO	Variability Analyses of Surface Climate Observations

Chapter 1
Introduction

1.1 Introduction

The global demand for food is rising steeply as a result of burgeoning population, shifting dietary preferences, and food wastage, while increasing demands for renewable energy are competing with food production (Hubert et al. 2010). In 2009, the FAO estimated that we must increase the global food production by 70 % to meet demands in 2050 (FAO 2009). But this figure is questioned and may be an underestimate, which further underlines the urgency of global food provisioning (Tilman 2010; Tilman et al. 2002), particularly in the light of the revised World Population Prospects 2012 predicting significantly higher population increase than earlier projections, especially for many countries in sub-Saharan Africa (UN 2013). Further, accelerating climate change is projected to have severe impacts on crop productivity over large parts of the globe (Porter et al. 2014). The combination of increasing water scarcity, as a result of climate warming, and increasing competition across sectors is likely to cause dramatic situations in terms of food and water security in many regions (Strzepek and Boehlert 2010). At the same time "business as usual is not an option." This was the stern message from the International Assessment of Agricultural Science and Technology (IAASTD) when it was presented by its chairman Bob Watson in 2008. By this he meant that agriculture does not deliver what we need—food security for all—instead it undermines the global environment in terms of land degradation; greenhouse gas emissions; pollution of soils, rivers, lakes, and oceans; and reducing biodiversity (Foley et al. 2011). The threat to food security represents a planetary emergency that demands a variety of creative solutions and policies at global, regional, and local levels. One of the most urgent responses to this situation must be measures to stop and reverse land degradation. But such solutions are currently hampered by the lack of reliable data as well as methods for collecting such data. This report is a review of the state of the art of remote sensing techniques for assessing land degradation and improvements.

© The Author(s) 2015
G.T. Yengoh et al., *Use of the Normalized Difference Vegetation Index (NDVI) to Assess Land Degradation at Multiple Scales*, SpringerBriefs in Environmental Science, DOI 10.1007/978-3-319-24112-8_1

1.2   Land Degradation in the UNCCD and GEF

Land degradation has been highlighted as a key development challenge by the UNCCD, the Convention on Biodiversity, the Kyoto Protocol on global climate change, and the Millennium Development Goals (United Nations 2011; UNEP 2007). The GEF was designated a financial mechanism for the UNCCD in 2003; through establishment of its land degradation focal area, the GEF aims to arrest land degradation, especially desertification and deforestation, by supporting sustainable land management (SLM). SLM implements agricultural practices that maintain vegetative cover; build up soil organic matter; make efficient use of inputs such as water, nutrients, and pesticides; and minimize off-site impacts (Bierman et al. 2014).

Both the UNCCD and the GEF use land cover to monitor land degradation and implementation of SLM. Likewise, the trend in land cover is a key indicator of progress in meeting the UNCCD's Strategic Objective 2: to improve the condition of affected ecosystems (UNCCD decision 22/COP.11). For the GEF, achievement of the overall goal of the land degradation focal area is measured through *"change in land productivity"* using, as a proxy, net primary productivity NPP which is estimated through remotely sensed normalized difference vegetation index (NDVI) screened for drought effects using rain-use efficiency RUE. To measure the impact of interventions, GEF-funded SLM projects should report on changes in land cover (GEF 2014). The same approach has also been used to allocate resources from the land degradation focal area of the GEF; other things being equal, countries suffering from serious land degradation, as measured as change in NDVI, are allocated more funds than those with lesser measurable evidence of land degradation.

Recent improvements and the longer time series of the fundamental NDVI dataset call for a review of indicators for measuring the implementation of the Convention and the GEF's allocation of resources to combat land degradation, as well as for measuring the impacts of its SLM projects.

1.3   Concepts, Processes, and Scales of Land Degradation

Land is defined as the *"ensemble of the soil constituents, the biotic components in and on it, as well as its landscape setting and climatic attributes"* (Vlek et al. 2010). Land degradation is a composite concept that has been defined in many and various ways. Indeed, it is a concept as much as a process, defined in various ways by researchers and institutions in this field. This could partly be as a result of the diversity of processes of land degradation in type, scale, time, and extent; the processes are well known but not always fully understood. According to Warren (2002), land degradation is a very contextual phenomenon and cannot *"be judged independently of its spatial, temporal, economic, environmental and cultural context."* This ambiguity makes it hard to establish measurable indicators, remotely sensed or otherwise.

Stocking and Murnaghan (2000) describe land degradation as a composite term that *has no single readily-identifiable feature, but instead describes how one or more of the land resources (soil, water, vegetation, rocks, air, climate, and relief) has changed for the worse* (Fig. 1.1). Haigh (2002) offers a more utilitarian definition: *the aggregate diminution of the productive potential of the land, including its major uses (rain-fed, arable, irrigated, rangeland, forest), its farming systems (e.g., small-holder subsistence) and its value as an economic resource.* This definition highlights deterioration in the biological productive potential of the land, i.e., the entire geo-ecological system that includes soils, climate, biodiversity, topography, and land use. The key message conveyed by this definition is akin to that conveyed by the Millennium Ecosystem Assessment's definition of land degradation, *the reduction in the capacity of the land to perform ecosystem goods, functions and services that support society and development* (MEA 2005). According to UNEP (2007), *land degradation is the long-term loss of ecosystem function and services, caused by disturbances from which the system cannot recover unaided.* This definition conveys two important messages: the resilient properties of landscapes and their constituent parts and the need for intervention if and when disturbances cause the resilience thresholds to be breached.

Degradation may also be considered in terms of specific components of the land that are affected. For example, vegetation degradation implies reduction in productivity, declining species diversity, and degeneration in the nutritional value of plant populations for the faunal biota. And soil degradation implies deterioration in soil quality and fertility. Such changes may be brought about by many factors (erosion, pollution, deforestation, and others). Again, land degradation may be considered in respect of its physical aspects, referring to changes in the soil composition, especially loss of soil organic matter, and structure, such as compaction or crusting and waterlogging; chemical, pertaining to changes in the soil chemistry's chemical makeup as a result of leaching, salinization, or acidification; and biological degradation referring to reduction of soil biodiversity.

Estimates of the extent and severity of land degradation vary substantially. The only agreement has been that all global estimates have rested on very poor data (Hassan et al. 2005). The Millennium Ecosystem Assessment reported estimates

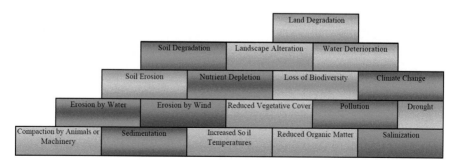

Fig. 1.1 The complexity of processes that constitute land degradation (Stocking and Murnaghan 2000)

between 70 and 10 % of drylands globally being affected by land degradation and concluded with *medium certainty that some 10–20 % of the drylands are suffering from one or more forms of land degradation. And the livelihoods of millions of people ... are affected* (Hassan et al. 2005). These figures were, however, not based on a systematic assessment of empirical data.

In order to overcome this uncertainty barrier, GEF/UNEP/FAO initiated the LADA project (Land Degradation in Drylands) which adopted the approach used by Bai and others (Bai et al. 2008). Based on the analysis of a 30-year time series of global NDVI data in combination with gridded climate data, Bai et al. (2008) reported that about 20 % of cultivated land, 30 % of forests, and 10 % of grasslands are degrading. Many studies have reported increasing severity and extent of land degradation in many parts of the world, but estimates tend to be highly method specific (see Annexes 1 and 2).

Land degradation can be caused by local human activities and biophysical processes as well as by activities and processes that are not tied to the local human or physical landscape (Fig. 1.2). Local activities that contribute to land degradation include mining, unsustainable farming practices, overgrazing, pollution from industrial and nonindustrial sources, and landscape modification. Hoekstra et al. (2005)

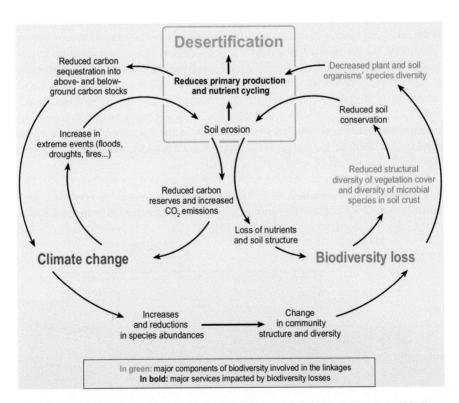

Fig. 1.2 Linkages and feedback loops among desertification, global climate change, and biodiversity loss (MEA 2005)

argue that land degradation resulting from human conversion of natural habitats is most extensive in tropical dry forests (69 % converted in SE Asia), temperate broad-leaf and mixed forests, temperate grasslands and savannas (>50 % lost in North America), and Mediterranean forest and scrub. Human activities responsible for land degradation go beyond farming practices, deforestation, and other direct human interactions with the land (Hoekstra et al. 2005). UNEP (2012a) and MEA (2005) see the causes of desertification (nefarious land degradation affecting people in arid and semiarid regions) ranging from international economic activities to unsustainable land-use practices by local communities. It has also been argued that processes such as dryland degradation may be exacerbated by climate change (Cowie et al. 2011).

1.4 Assessment of Resilience of Agroecosystems

No less than land degradation, resilience is an ambiguous term (Thorén and Persson 2014) subject to scientific and political debates (Walker et al. 2004). In his seminal paper in 1973, Holling writes: *Resilience determines the persistence of relationships within a system and is a measure of the ability of these systems to absorb change of state variable, driving variables, and parameters, and still persist* (Holling 1973). Perrings (1998) offered a more open definition: *in its broadest sense, resilience is a measure of the ability of a system to withstand stresses and shocks—its ability to persist in an uncertain world*, and interdisciplinary scientists interested in coupled social and ecological systems (SESs) have incorporated the idea into their thinking, as expressed by Adger: *The ability of human communities to withstand external shocks or perturbations to their infrastructure, such as environmental variability or social, economic or political upheaval, and to recover from such perturbations* (Adger 2000).

Renschler et al. (2010) have argued that environmental and ecosystem resources might be used as indicators of ability of the ecological system to return to or near pre-shock or pre-event states. The strong correlation of NDVI with aboveground NPP makes it a useful indicator of ecosystem resilience. In a study exploring the concepts and application of theories of general resilience, Walker et al. (2014) identified twelve components of general resilience in five catchments in south eastern Australia. These components include diversity (which may be identified and measured by processes including vegetation clearing, forest fires, floods, and drought), and connectivity, modularity, and reserves in ecological systems (Walker et al. 2014) which can be identified and measured by earth observation methods, including land-use and land-cover change assessments. In the context of monitoring land degradation using remotely sensed data, we would prefer a more precise definition of resilience that can be operationalized by something measurable. A central concept in ecological resilience is a system's ability to absorb and recover from disturbance or stress; this may be depicted by a hysteresis curve (Kinzig et al. 2006) (Fig. 1.3).

A resilient system subject to stress, such as drought, may reduce its productivity as long as the stress persists but, then, return to its prestress productivity. If the system is not sufficiently resilient, it will not regain its prestress productivity. The Sahel is an example of resilience at a grand scale. Since the 1980s, long time series of NDVI data have been used extensively in the study of land degradation in the Sahel (Fensholt et al. 2013; Anyamba and Tucker 2005; Hickler et al. 2005; Prince et al. 1998), confirming a general pattern of recovering vegetation.

The interpretation of the recovery of vegetation vis-à-vis the resilience of such systems must, however, be approached with caution. This is because the state of an ecosystem is not defined solely by its overall bio-productivity, but also, by the vegetation composition as well as the ecosystem services it offers. It therefore follows that the stability of positive trends in bio-productivity (an aspect of ecosystem dynamics that can be captured by the time-series analysis of NDVI data) may not necessarily report the resilience of such systems. Recent studies relating long-term NDVI trends to ground observations in Senegal show that positive NDVI trends do not systematically indicate positive developments, neither in terms of the composition of the vegetation cover, which showed impoverishment even in the greening areas (Herrmann and Tappan 2013), nor in terms of human well-being (Herrmann et al. 2014).

NDVI is proposed as a measure of *land-cover status*—one of the eleven impact indicators recommended in the UNCCD "Minimum set of Impact Indicators"; its purpose (Orr 2011) is to *monitor land degradation in terms of long-term loss of ecosystem primary productivity and taking into account effects of rainfall on NPP*.

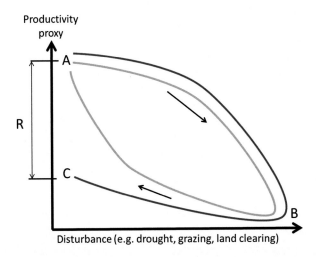

Fig. 1.3 The principle of hysteresis. At point *A*, before the stress, productivity is high. As the stress increases, productivity declines to a point *B* where the stress is reduced. As the stress is reduced, productivity increases. A fully resilient system (*green curve*) will spring back to its original state (*A*). A less-resilient system (*red curve*) will only recover to point *C*. The resilience of the system, *R*, is related to the distance between *A* and *C*; the lower the value, the higher the resilience

DPSIR (Driving Force, Pressure, State, Impact, Response) is a general framework for organizing information and reporting about state of the environment. First developed by the Organization for Economic Cooperation and Development (OECD) in the 1980s, this framework is currently being applied in a range of fields and projects, including those of the UNCCD and GEF (Orr 2011). DPSIR is also the methodological framework used by UNEP in its Global Environment Outlook (GEO) reports at global, regional, and national levels (UNEP 2012a). The state variables are pointers to the condition of the system (including biophysical factors/processes), as well as trends (environmental changes) which may be naturally or human induced (Vacik et al. 2007; Orr 2011). NDVI can be useful in the evaluation of vegetation cover, carbon stocks, and land condition (Orr 2014) which may provide resilience indicators.

Chapter 2
The Potential for Assessment of Land Degradation by Remote Sensing

Given the diversity of the biophysical and socioeconomic processes involved, the types, extent, and severity of land degradation cannot be encapsulated by a few simple measures (Stocking and Murnaghan 2000). In the assessment of land degradation or changes in land productivity, two complementary approaches may be distinguished:

1. An assessment of historic trends in land degradation or changes in land productivity, in which past changes are examined
2. An assessment of future trends, in which scenario building and projections are made of expected changes in land degradation or land productivity based on defined scenarios

For a comprehensive assessment, monitoring, and mapping of land degradation, four main themes need to be explored:

1. Causes of degradation—the drivers, mostly man-made such as agricultural practices, overgrazing, deforestation, and industrial activities such as mining
2. Type of degradation—the nature of the process driving decline in land quality or productivity. For example, drought, salinization, and wind or water erosion
3. Degree of degradation—classified in degrees of severity, such as light, moderate, strong, and extreme
4. Extent of degradation—the total area affected

2.1 Normalized Difference Vegetation Index

The last half century has seen the development and use of various remotely sensed vegetation indices. The basic assumption behind the development and use of these indices is that some algebraic combination of remotely sensed spectral bands can reveal valuable information such as vegetation structure, state of vegetation cover, photosynthetic capacity, leaf density and distribution, water content in leaves,

© The Author(s) 2015
G.T. Yengoh et al., *Use of the Normalized Difference Vegetation Index
(NDVI) to Assess Land Degradation at Multiple Scales*, SpringerBriefs
in Environmental Science, DOI 10.1007/978-3-319-24112-8_2

mineral deficiencies, and evidence of parasitic shocks or attacks (Jensen 2007; Liang 2005). The algebraic combination of spectral bands should, therefore, be sensitive to one or more of these factors. Conversely, a good vegetation index should be less sensitive to factors that affect spectral reflectance such as soil properties, atmospheric conditions, solar illumination, and sensor viewing geometry (Jensen 2007; Liang 2005; Purkis and Klemas 2011).

The structure of leaves, evolved for photosynthesis, determines how vegetation interacts with sunlight. Two processes occur within leaves: absorption and scattering of sunlight. Plant pigments (chlorophyll and carotenoids) and liquid water absorb specific wavelengths of light. Scattering is caused by the internal structure of leaves, where the leaf interior is a labyrinth of air spaces and irregularly shaped water-filled cells. Internal scattering of light is caused by differences in the refractive index between air- and water-filled cells and internal reflections from irregularly shaped cells. Green leaves absorb light strongly in the blue and red regions and less so in the green region, hence their green color (Jensen 2007). No absorption occurs from the upper limit of our vision at 700 nm out to beyond 1300 nm where liquid water begins to absorb strongly (Fig. 2.1). No absorption means higher levels of reflectance from green vegetation (Tucker and Garratt 1977).

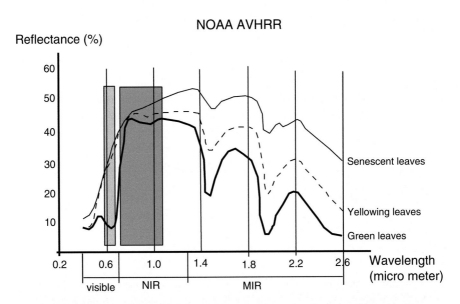

Fig. 2.1 Spectral response characteristics of vegetation at three stages of development. The spectral bands of the most commonly used sensor for NDVI studies, NOAA AVHRR, are superimposed on the spectral response curve. Chlorophyll contained in a leaf has strong absorption at 0.45 and 0.67 μm and high reflectance in the near-infrared (0.7–1.1 μm). In the shortwave-IR, vegetation displays three absorption features that can be related directly to the absorption of water contained within the leaf

The normalized difference vegetation index (NDVI) (Eq. 2.1) is the ratio of the difference between the near-infrared band (NIR) and the red band (R) and the sum of these two bands (Rouse Jr et al. 1974):

$$NDVI = \frac{NIR - RED}{NIR + RED} \qquad (2.1)$$

where NIR is reflectance in the NIR and RED is reflectance in the visible red band. The NDVI algorithm takes advantage of the fact that green vegetation reflects less visible light and more NIR, while sparse or less green vegetation reflects a greater portion of the visible and less near-IR. NDVI combines these reflectance characteristics in a ratio so it is an index related to photosynthetic capacity. The range of values obtained is between −1 and +1. Only positive values correspond to vegetated zones; the higher the index, the greater the chlorophyll content of the target.

NDVI has been used to identify and interpret a range of phenology metrics that describe periodic plant life-cycle events and how these are influenced by seasonal and interannual variations in climate and habitat (see Annexes 1 and 2). So the duration of photosynthetic activity (identified using NDVI) can be interpreted to indicate the length of the growing season; time of maximum NDVI corresponds to time of maximum photosynthesis; seasonally integrated NDVI indicates photosynthetic activity during the growing season; and the rate of change in NDVI may indicate speed of increase or decrease of photosynthesis. These metrics are influenced by several characteristics of the vegetation. One of the most important in remote sensing is the leaf area index (LAI) which refers to the projected area of leaves per unit of ground area (Ross 1981).

2.2 Remote Sensing Features That Characterize NDVI-Based Assessments of Land Degradation

Potential for the use of NDVI as a proxy for land productivity (one of the indicators of the state of land degradation) is based on numerous and rigorous studies that have identified a strong relationship between NDVI and NPP (Prince and Goward 1995; Vlek et al. 2010; Field et al. 1995) (also see Sect. 5.5). Remotely sensed data products derived from satellite measurements come in several bands of the electromagnetic spectrum (see Annex 5). NDVI and related indices use bands in the visible and infrared wavelengths. When using satellite-derived products, it is important to consider sensor and image characteristics such as image size, region of the earth from which images are acquired, spatial resolution, number of bands and wavelengths detected, spectral characteristics of the bands concerned, frequency of image acquisition, and date of origin of the sensor (Strand et al. 2007). Another important consideration is the time of acquisition of such data (time of the day, or season in question). Such temporal differences may give rise to alterations such as shadows (depending on the time of the day) or phenological differences (depending on the season) that may affect the quality of the data. Remote sensing products rarely meet all requirements for image size, spatial and temporal resolution, and availability. There is always a need for trade-offs (Purkis and Klemas 2011; Strand et al. 2007). Images with large path width have low spatial resolution, lower data volume, and shorter temporal resolutions—so they tend to have a longer time series from which long-term changes can be observed. With the large path width of low-resolution imagery, large areas can be covered and analyzed by a few images. On the other hand, high spatial resolution is associated with a smaller path width, large data volumes, and longer temporal resolution; this demands greater resources in data storage, manipulation, and analysis. Also, most high-resolution datasets are pricey—beyond the reach of many potential users outside the research community of the satellite launching program or country (see Annex 7 on current costs of some satellite data products). In general, high spatial resolution data are helpful for fine-scale assessments and analysis at local level, while medium spatial resolution data are useful at a regional or, even, project scale. At a continental or global scale, coarse spatial resolution data support archives of long time series and are preferred for many NDVI-based assessments and analyses. Long time series simplify the use of remote sensing to assess land degradation and monitor changes (Albalawi and Kumar 2013; Anyamba and Tucker 2012; Bai et al. 2008; Cook and Pau 2013; de Jong et al. 2011b; Shalaby and Tateishi 2007; Symeonakis and Drake 2004; Townshend et al. 2012).

2.3   Other Vegetation Indices Closely Related to NDVI

One of the earliest attempts at separating green vegetation from the soil background using the NIR/red ratio was carried out by Pearson and Miller (1972). Since then, many and various vegetation indices have been developed, tested, modified, and used for vegetation-related studies worldwide. These include LAI, percent vegetation cover, green leaf biomass, fraction of absorbed photosynthetically active radiation (fAPAR), photosynthetic capacity, and carbon dioxide fluxes (Albalawi and Kumar 2013; Liang 2005; Purkis and Klemas 2011). More than 150 vegetation indices have appeared in the literature although few have been systematically tested (Bennett et al. 2012; Verrelst et al. 2006; Higginbottom and Symeonakis 2014). Vegetation indices derived from satellite data are one of the principal sources of information for monitoring and assessment of the Earth's vegetative cover (Gilabert et al. 2002). We direct readers to these references for background information on vegetation indices and focus our attention on the NDVI and related vegetation indices.

2.3.1   Indices Closely Related to NDVI

The *Enhanced Vegetation Index* (EVI) (Eq. 2.2) is a modification of NDVI with a soil adjustment factor, L, and two coefficients, C_1 and C_2, which describe the use of the blue band in correction of the red band for atmospheric aerosol scattering. C_1,

C_2, and L, are coefficients that have been empirically determined as 6.0, 7.5, and 1.0, respectively. EVI was developed by the MODIS Land Discipline Group for use with MODIS data to decouple the canopy background signal and reduce atmospheric influences (Huete et al. 2002; Jensen 2007). However, subsequent work with the EVI has raised two scientific controversies resulting from use of this index (Morton et al. 2014; Saleska et al. 2007). Because of this controversy, NDVI is always preferred over EVI. Furthermore, it has been difficult to inter-calibrate EVI between or among different instruments because the surface reflectance uncertainty of the blue band is high for dense green vegetated areas. Eric Vermote (Personal communication 2014) has found the MODIS blue surface reflectance from dense green vegetation to be on the order of 3–4 % with an absolute surface reflectance uncertainty of ±2 - ±3 %. For this reason, the VIIRS vegetation index science team has proposed discontinuing the three-channel EVI and replacing it with a substitute two-channel EVI that is a modified NDVI (Eqs. 2.3 and 2.4). This was first proposed by Jiang et al. (2008):

$$ EVI = 2.5 \times \frac{(NIR - RED)}{(NIR + C_1 \times RED - C_2 \times BLUE + L} \tag{2.2} $$

2.3.2   Comparing NDVI to EVI

Here, we discuss only those vegetation indices for which current and freely available global datasets exist. The contenders are NDVI, the three-channel EVI of Huete et al. (2002), and the two-channel EVI of Jiang et al. (2008).

The three-channel EVI will be discontinued for VIIRS because of problems with the calibration of the blue band (blue surface reflectance from dense green vegetation is ±2–3 %); problems with sub-pixel clouds, aerosols, and snow; the fact that MODIS data are atmospherically corrected; and the realization that the blue band is very highly correlated to the red band for vegetation. The three-channel EVI has been replaced by a two-channel EVI (Jiang et al. 2008). Figure 2.2 shows the three-channel EVI sensitivity to aerosols, smoke, sub-pixel clouds, and snow. It also shows the very high correlation between the blue and red surface reflectances from vegetated areas which, therefore, adds no new information to this index.

Because of the problems with the EVI3's blue band, illustrated in Fig. 2.2a, the MODIS three-channel EVI products were frequently produced with a two-channel red and near-infrared "soil-adjusted vegetation index" or SAVI substitute algorithm, depending on circumstances. Thus, MODIS three-channel EVI data were, in practice, a combination of three-channel and two-channel products and not the same numerical product everywhere all the time. For these reasons, we propose to use the NDVI.

The two-channel EVI proposed by Jiang et al. (2008) (Eq. 2.3) is very similar to NDVI and is directly related to the NDVI and gives more weight to the NIR:

$$ EVI2 = 2.5 \times (NIR - Red) / (NIR + 2.4 \times Red + 1) \tag{2.3} $$

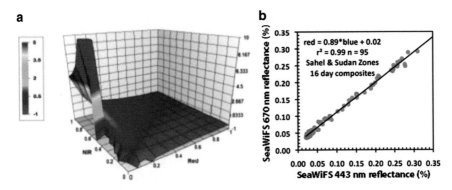

Fig. 2.2 The three-channel *blue, red,* near-infrared EVI vegetation index suffers from not being atmospherically resistant and is very sensitive to high *blue-band* surface reflectance. (**a**) Panel shows the erroneous three-channel EVI values from sub-pixel clouds, smoke, aerosols, and snow. (**b**) Panel shows the very high correlation between the *blue* and *red bands* for vegetated areas. Information theory tells us that highly correlated variables do not increase the variance explained together over using just one of the variables

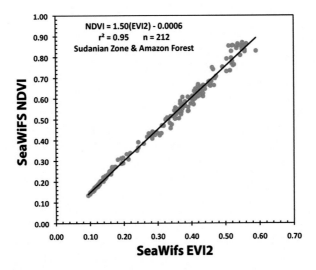

Fig. 2.3 A NDVI and EVI2 comparison using SeaWiFS data from the Sudanian Zone of Africa combined with similar data from the Central Amazon for 212 points. Note the very high degree of similarity between the NDVI and the two-channel EVI

What then is the advantage of the EVI2 over the NDVI? Multiplying each side of Eq. 2.3 by (NIR + RED)/(NIR + RED) or 1.0 and rearranging terms gives Eq. 2.4:

$$\text{EVI2} = \text{NDVI} \times 2.5 \times (\text{NIR} - \text{Red}) / (\text{NIR} + 2.4 \times \text{Red} + 1) \qquad (2.4)$$

The NDVI and EVI2 are very similar, with the NDVI being directly related to primary production, and the EVI2 being more heavily weighted to mapping LAI in very dense plant canopies (Fig. 2.3).

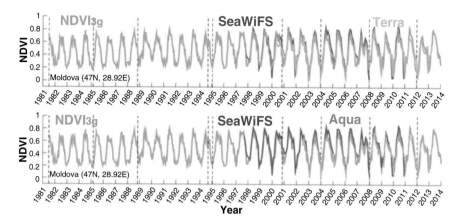

Fig. 2.4 Comparison of NDVI data from eight different AVHRR instruments at nine times from 1981 to 2014 (noted by the *vertical dashed lines*) to NDVIs from SeaWiFS during 1997–2010, Terra MODIS from 2000 to 2013, and Aqua MODIS from 2002 to 2013. T NDVI time series from these four sources are very similar. There are currently three AVHRR instruments operating, two with MetOps and one with NOAA, and one in storage awaiting launch on MetOps-3 in 2016. There is an excellent chance that we will have a 35–40-year NDVI record from the AVHRR instruments

Quantitative inter-comparability between or among similar satellite instruments is important because a global analysis at 8 km spatial resolution, using data over 33 years, can identify specific areas of possible land degradation which then can be investigated in more detail with much higher spatial resolution time-series data from MODIS at 250 m. Fortunately, NDVI with only two channels lends itself to quantitative inter-comparability among similar satellite instruments (Fig. 2.4).

The VIIRS instrument now flying on the NASA-NOAA *NPOESS Preparatory Project* polar-orbiting meteorological satellite will continue to deliver NDVI and two-channel EVI data through 2030 and beyond on the NASA-NOAA Joint Polar Satellite System (JPSS) five satellites (JPSS-1 through JPSS-5). The three-channel EVI will be discontinued.

We conclude that NDVI has fewer problems than the three-channel EVI because it can be inter-calibrated more easily with only two bands and will be replaced by the EVI2 from the NPP and JPSS-1 to JPSS-5 satellites. We choose to use the standard NDVI to identify land degradation because it is directly related to photosynthetic capacity. However, when EVI2 data are available, we and others will evaluate this vegetation index for land degradation also. Excellent NDVI data are available from MODIS, and these form the foundation of our work because we require 250 m NDVI data for disaggregation (Huete et al. 2002). We stress that all of our proposed land degradation NDVI work must also use MODIS NDVI data from 2000 to 2014 to confirm 8 km NDVI3g data from the same time period.

Chapter 3
Applications of NDVI for Land Degradation Assessment

In the late 1960s, several researchers began using red and near-infrared reflected light to study vegetation (Pearson and Miller 1972). In the late 1960s, ratios of red and near-infrared light were used to assess turf grass condition and tropical rain forest leaf area index (Birth and McVey 1968; Jordan 1969). Compton Tucker was the first to use it for determining total dry matter accumulation, first from hand-held instruments (Tucker 1979), and then from NOAA AVHRR satellite data (Tucker et al. 1981, 1985), demonstrating that the growing season integral of frequent NDVI measurements represented the summation of photosynthetic potential as total dry matter accumulation. Starting in July 1981, a continuous time series of global NDVI data at a spatial resolution of 8 km has been available from the AVHRR instrument mounted on NOAA weather satellites. Soon, researchers realized the value of NDVI time-series remote sensing (Goward et al. 1985; Justice et al. 1985; Townshend et al. 1985; Tucker et al. 1985). This early work was the spur for development of the higher-resolution Moderate-Resolution Imaging Spectroradiometer (MODIS) instrument. The application of satellite NDVI data has blossomed into many fields of natural resources investigation (see Annex 1). One particular appeal of remote sensing in the study of large geographic areas, or at multiple times over the year(s), is the potential for cost savings (Pettorelli 2013). We examine the use of NDVI in research on land-use and land-cover change, drought, desertification, soil erosion, vegetation fires, biodiversity monitoring and conservation, and soil organic carbon (SOC).

3.1 Land-Use and Land-Cover Change

Land cover is the observed (bio)physical cover on the earth's surface (Di Gregorio 2005). In a strict sense, it should be limited to the description of vegetation and artificial (man-made) features on the land surface, but land-use and land-cover change (LULCC) is a general term used to refer to the human modification of the

© The Author(s) 2015
G.T. Yengoh et al., *Use of the Normalized Difference Vegetation Index (NDVI) to Assess Land Degradation at Multiple Scales*, SpringerBriefs in Environmental Science, DOI 10.1007/978-3-319-24112-8_3

earth's terrestrial surface (Bajocco et al. 2012). Mankind has modified land and land cover for thousands of years to obtain food, fuel, fiber, and other materials, but current rates and intensities of LULCC are far greater than ever before (Lambin et al. 2003; Mayaux et al. 2008). The quantity and quality of vegetation cover are an important controls on the evolution landscapes, their resilience or degradation (Symeonakis and Drake 2004), and the quality of environmental services.

Substantial research effort has been invested in the use of NDVI to assess the location and extent of land-use and land-cover change (Diouf and Lambin 2001; Mas 1999; Stow et al. 2004; UNEP 2012b; Veldkamp and Lambin 2001; Yuan and Elvidge 1998). Land-cover change analysis is used to study changes in plant composition and human settlement patterns (Khorram et al. 2012). Such studies are important in understanding the scale and reasons for changes in vegetation, biodiversity, and associated phenomena. Scales of application range from global studies of land-cover classification and mapping (DeFries and Townshend 1994; Friedl et al. 2002; Hansen et al. 2000; Turner and Meyer 1994), to regional and national scales (Lambin and Ehrlich 1997; Sobrino and Raissouni 2000; Stow et al. 2004), to very localized studies (Shalaby and Tateishi 2007; Sternberg et al. 2011; Yuan and Elvidge 1998; Lunetta et al. 2006). Horion et al. (2014) used long-term trends in dry-season minimum NDVI to assess changes in tree cover in the Sahel; dry-season minimum NDVI was found to be uncorrelated with dry grass residues from the preceding growing season or with seasonal fire frequency and timing over most of the Sahel, so the NDVI parameter can be used as a proxy for assessing changes in tree cover in such ecosystems. While land-use and land-cover change can serve as pointers to the existence (or absence) of land degradation, care must be taken in interpreting the results of such studies. Veldkamp and Lambin (2001) warn about the need to distinguish between the location and quantity of change, as well as the causes of such changes. For example, while NDVI can help in identifying deforestation, its rate, and area affected, the underlying drivers are often far away in space and time (Veldkamp and Lambin 2001). And there are many cases where land-cover change as detected by NDVI time series does not necessarily lead to degradation but may rather be considered beneficial (Lambin et al. 2003; Shalaby and Tateishi 2007).

3.2 Drought and Drought Early Warning

Drought generally refers to a substantial decline in the amount of precipitation received over a prolonged period (Mishra and Singh 2010). Droughts occur in practically all climatic zones and are recognized as a severe hazard to the environment and development (Sivakumar and Stefanski 2007). In terms of land degradation, droughts cause loss of water availability and quality, declining primary production which increases the vulnerability of the land to erosion and disturbed riparian habitats, with potential loss of biodiversity (Mishra and Singh 2010; Zargar et al. 2011).

NDVI and associated vegetation indices have been used to detect and investigate meteorological, hydrological, and agricultural droughts worldwide. Strictly speaking,

NDVI is most useful for detecting and investigating drought effects on the vegetation cover, in this case agricultural droughts. Generally, meteorological (dry weather patterns) and hydrological (low water supply) droughts would not be detected by NDVI before they impact the vegetation cover. The exception in this case may be hydrological drought, if the analysis includes water level in plant matter. NDVI has been used by several studies of the Sahelian drought (see Annex 1) an example is the investigation of the persistence of drought in the Sahel in the period 1982–1993 (Anyamba and Tucker 2012). Another study combined anomalies of El Niño Southern Oscillation (ENSO) indices and NDVI anomalies to construct an ENSO-induced drought onset prediction model for northeast Brazil using multiple linear regression (Liu and Juárez 2001). The normalized difference water index (NDWI) is a sister index calculated from the 500-m SWIR band of MODIS that has been used in drought studies (Chen et al. 2005; Delbart et al. 2005; Jackson et al. 2004; Gao 1996). Mishra and Singh (2010) argue that NDWI may be a more sensitive indicator than NDVI for drought monitoring, but its developer (Gao 1996) emphasized that the index is: *complementary to, not a substitute for NDVI.*

NDVI has also been used widely in attempts to develop famine early warning systems, such as the FEWS NET which is an operational system for dissemination of data related to food production and availability globally. The system uses NDVI data from both NOAA AVHRR and MODIS. The system was preceded by rigorous testing of the ability of NDVI to detect areas of imminent food shortages (Henricksen and Durkin 1986; Hutchinson 1991; Quarmby et al. 1993). A main finding was that NDVI in combination with relevant climate data has a very strong potential for forecasting crop failure.

3.3 Desertification

Beginning in the 1970s, the international community recognized that land degradation/desertification was an economic, social, and environmental problem and began a process which ultimately resulted in the creation of the UN Convention to Combat Desertification (CCD). The convention defines desertification as *land degradation in arid, semi-arid, and dry sub-humid areas resulting from various factors, including climatic variations and human activities* (UNCCD 1994). Some studies report that desertification poses a serious global threat, affecting both developed and developing countries (Grainger 2013): others report that drylands have been greening and caution against broad generalizations (Fensholt et al. 2012). Since the 1980s, remote sensing has been used extensively in the study of desertification in different parts of the world (Erian 2005; Fensholt et al. 2013; Karnieli and Dall'Olmo 2003; Nkonya et al. 2011; Symeonakis and Drake 2004; UNEP 2012b; Olsson et al. 2005; Tucker and Nicholson 1999). While studies in the 1980s demonstrated the value of the NDVI for tracking vegetation dynamics, a clear relationship between NDVI, biomass accumulation (especially in the Sahel), and the many variables which interact with them was not fully understood (Herrmann and Sop 2015). Most studies simply referred to vegetation trends and embed their hypotheses and findings in the larger debate on desertification. Beyond showing changes in bioproductivity, interpretation of those trends was rather speculative. Today, notwithstanding the significant advances in knowledge of these processes, desertification remains a difficult process to assess given the complex relationship between biomass and ecosystem health (Herrmann and Sop 2015). More recently, there have been a several studies linking remotely sensed trends to ground observations (Brandt et al. 2014; Dardel et al. 2013; Herrmann and Tappan 2013). Examples of such recent developments in the understanding of desertification also include the use of land productivity dynamics in constructing a World Atlas of Desertification (WAD) (see Annex 2).

In detecting the status and trend of desertification, researchers have built on the relationship between NDVI and biomass productivity that has been well established in the literature (Jensen 2007; Purkis and Klemas 2011). These initiatives are greatly helped by the continuous global NDVI time series of vegetation that has been available since the early 1980s. The UNCCD fostered increased interest in desertification research, especially with regard to the Sahel which was by that time experiencing a wet period, captured by satellite imagery, analyzed using NDVI time series, and described as a *greening of the Sahel* (Olsson et al. 2005). The following studies used NDVI time series to investigate temporal and spatial patterns of the Sahel's greenness and rainfall variability as well as their interrelationships (Herrmann et al. 2005; Hickler et al. 2005). Herrmann et al. (2005) and Olsson et al. (2005) concluded that while increased rainfall was the main reason for greening, there were also a number of hypothetical human-induced changes superimposed on the climatic trend-changes such as improved agricultural practices, as well as migration and population displacement. Besides documenting the close coupling of rainfall and vegetation response in the Sahel, Anyamba and Tucker (2005) pointed out that current greener conditions are still not as green as those that prevailed from 1930 to 1965. Together, these studies and many that have followed (see Annexes 1 and 2) demonstrate the opportunities offered by NDVI as a proxy for vegetation response to rainfall variability, especially in arid and semiarid ecosystems. Herrmann et al. (2005) also demonstrated the possibility of using NDVI as a proxy for environmental response to management.

3.4   Soil Erosion

Erosion is the displacement of materials like soil, mud, and rock by gravity, wind, water, or ice. The most common agents of soil erosion are water and wind (Foth 1991); their effects may be on-site (where soil detachment and transportation occurs) or off-site (where eroded soil is deposited). In soil erosion studies, NDVI is commonly used in conjunction with soil-erosion estimation models such as the fuzzy-based dynamic soil erosion model (FuDSEM), the Revised Universal Soil Loss Equation (USLE/RUSLE), the Water Erosion Prediction Project (WEPP), the European Soil Erosion Model (EUROSEM), and the Soil and Water Assessment

Tool (SWAT) (Prasannakumar et al. 2012; Zhou et al. 2008). Mulianga et al. (2013) in Kenya and Ai et al. (2013) in China, used Terra-MODIS-derived NDVI to characterize the state of the ecosystem (spatial and temporal heterogeneity of the vegetation conditions), and as one of the input parameters for estimating the potential of erosion using fuzzy-set theory. In a study of the effect of vegetation cover on soil erosion in the Upper Min River watershed in the Upper Yangtze Basin, China, Zhou et al. (2008) used NDVI as a land-management factor (an input into the RUSLE model) representing the effect of soil disturbing activities, land cover, and vegetation productivity on soil erosion. A similar study, also using NDVI as a land-cover management factor to determine the vulnerability to erosion of soils, was carried out in a forested mountainous sub-watershed in Kerala, India (Prasannakumar et al. 2012). In such studies, NDVI proved to be a useful indicator of land-cover condition and a reliable input into models of soil dynamics.

   In most soil erosion research, the NDVI data come from Landsat TM/ETM (Thematic Mapper/Enhanced Thematic Mapper) with a spatial resolution of 30 m (Ai et al. 2013; Chen et al. 2011) or MODIS with a spatial resolution of 250 m (Fu et al. 2011; Mulianga et al. 2013). These data are generally used in conjunction with a digital elevation model with a spatial resolution of 30 m (Ai et al. 2013; Fu et al. 2011; Mulianga et al. 2013; Prasannakumar et al. 2012).

3.5 Soil Salinization

Salinity is salt in the wrong place, affecting water quality, uptake of water, and nutrients by plants and breaking up roads and buildings. It occurs naturally in drylands and areas prone to tidewater flooding but is often exacerbated by poor soil and water management (Zinck and Metternicht 2008). A quarter of global cultivated land is saline and one third is sodic (high in adsorbed sodium), but salinity can vary significantly, even over short distances. While soil salinization is a global problem, the phenomenon is more extensive in dry regions than in humid ones (Zinck and Metternicht 2008). Salinity might be plain to see at the soil surface and has been mapped from air photos and Landsat visible-light imagery. However, most of the salt is deep in the regolith and it's hard to isolate the effects of soil salinity on vegetation from the effects of other factors (Lobell et al. 2010).

   Assessment employs a combination of methods including airborne and ground-based electromagnetic induction (EM), field sampling, and solute modeling (Farifteh et al. 2006; Dent 2007). Airborne EM measures salt in three dimensions to a depth of 300 m. Passive sensors like AVHRR and other NDVI methods cannot see below the surface, but this has not deterred users—e.g., Platonov et al. (2013) investigated whether the values of maximum multi-annual NDVI reflected the degree of soil salinity within agricultural areas of Syr Darya province of Uzbekistan. The study found that by calculating the maximum multi-annual NDVI values from satellite images, one can create more spatially detailed soil salinity maps using two methods: the pixel based and the average for fields (Platonov et al. 2013). NDVI was reported to be one of the best band combinations estimating for soil salinity for some crops, such as alfalfa and corn, but not for others such as cantaloupe or wheat (Eldeiry and Garcia 2010). On the other hand, in a study that evaluated the use of multi-year MODIS imagery in conjunction with direct soil sampling to assess and map soil salinity at a regional scale in North Dakota and Minnesota, Lobell et al. (2010) found that the Enhanced Vegetation Index (EVI) for a 7-year period outperformed the NDVI in showing a strong relationship with soil salinity. Nonetheless, the NDVI has been used to study top-soil salinity conditions at the local and regional scales in different geographical settings. One may doubt whether such studies can be generalized when surface occurrences are so variable from season to season and year to year and vegetation effects are so variable (Metternicht and Zinck 2003). Nearly all the salt—and salt movement—is deep underground: this is a case where NDVI is not a suitable technique.

3.6 Vegetation Burning

Fire is a common occurrence in many parts of the world. Wild fires in forests, savannas, mountain regions, and other ecosystems are integral to the evolution of some of these systems; periodic fires are important in maintaining many grassland, shrub steppe, and savanna ecosystems (Mitchell and Roundtable 2010). On the other hand, fires (wild or man-made) give rise to soil erosion, greenhouse gas emissions, soot, and bad air quality and diminish biodiversity, soil water retention capacity, and soil structure (Purkis and Klemas 2011). Within the latter context, fires may constitute a form of land degradation.

Satellite remote sensing has been used for modeling and mapping a variety of ecosystem conditions associated with fire risks, potential, and management. These include fire-fuel mapping risk estimation (De Angelis et al. 2012), fire detection, postfire severity mapping, and ecosystem recovery from fire stress. Besides the detection of active fires, remote sensing is also used to assess and quantify the spatial and temporal variations of changes in vegetation cover in fire-affected areas. Here, prefire and postfire images are essential for estimating areas affected, especially when supported by techniques of supervised classification. Lanorte et al. (2014) used NDVI time series to monitor vegetation recovery after disturbance by fire at two test sites in Spain and Greece. A similar study by Leon et al. (2012) used MODIS NDVI to monitor postfire vegetation response in New Mexico. Their study outlined the potential of using NDVI to monitor the recovery of vegetation cover after fire disturbance (Leon et al. 2012). NDVI is also used to determine phenological information of the area affected (Chuvieco et al. 2004; Díaz-Delgado et al. 2003). In pre-burn analysis, this information can be used to calculate and map fire-fuel availability as well as potential economic and ecological losses likely to result from such fires. In post-burn analysis, NDVI can be used in fire severity mapping to assess ecological recovery from fires (Malak and Pausas 2006; Díaz-Delgado et al. 2003), to estimate carbon emissions resulting from the fire episode, and to assess environmental impact (Isaev et al. 2002).

Besides NDVI, studies of vegetation fires make use of related vegetation indices such as the Modified Soil-Adjusted Vegetation Index (MSAVI), Enhanced Vegetation Index (EVI), and the Foliar Moisture Index (FMI) (Wang et al. 2010) and, also, indices specific to this field such as the Normalized Burn Ratio (NBR), NBR Change Index (dNBR), and the Normalized Thermal Index (NTI) (Wang et al. 2010). Vegetation fire research makes use of an array of satellite sensors: MODIS, ASTER, Advanced Land Imager (ALI), AVHRR, Landsat 5 TM, Landsat 7 ETM+, Spot 4 and 5, Quickbird-2, and IKONOS-2. For most tasks involving the use of NDVI such as fuel mapping, vegetation classification, and postfire burn area and severity assessment and mapping, the main sensors are AVHRR, MODIS, Landsat 5 TM, and Landsat 7 ETM+ (see Annex 1).

3.7 Soil Organic Carbon (SOC)

The absorption of carbon by soils and vegetation ecosystems goes some way towards offsetting worldwide fossil fuel emissions (Mishra and Singh 2010; Piao et al. 2009; Bernoux and Chevallier 2014). SOC is also increasingly recognized as an excellent indicator of the status and functioning of soils—hence progressively recommended in various international initiatives for monitoring soil quality (Bernoux and Chevallier 2014). In the 1980s and 1990s, for example, 10–60 % global carbon emissions were offset through this process. While the global pattern and sources of sinks are imperfectly understood (Piao et al. 2009), it is well accepted that vegetation plays an important role in carbon sequestration.

Vegetation data based on NDVI have been instrumental in the assessment and monitoring of key global biomes (see Annex 1). The first pan-tropical biomass map was developed through MODIS-GLAS data fusion in 2011 (Saatchi et al. 2011). This has been a benchmark for the assessment of biomass carbon stocks in support of REDD assessments at both project and national scales. NDVI has been used in regional studies of SOC. Together with terrain attributes and data on climate, land-use, and bedrock geology, NDVI data were used to predict the SOC pool for seven states in the Midwestern United States (Mishra et al. 2010). In another study, NDVI data were used as an input into the Carnegie–Ames–Stanford Approach (CASA) terrestrial ecosystem model to estimate losses of SOC resulting from wind erosion in China (Yan et al. 2005). In investigating changes in soil organic C and total N in the Hexi Corridor, China, significant correlation was found between NDVI and SOC, as well as NDVI and N (Pan et al. 2013).

3.8 Biodiversity Monitoring and Conservation

High rates of biodiversity loss threaten to breach planetary boundaries (Rockström et al. 2009). A growing body of knowledge is developing around the tools and techniques for assessing and predicting ecosystem responses to global environmental changes (Pettorelli et al. 2005, 2014; Yeqiao 2011). In mapping and studying protected lands, for example, Yeqiao (2011) notes that satellite remote sensing can provide wide-ranging geospatial information at various spatial scales, temporal frequencies, spectral properties, and spatial contexts. While traditional approaches to measuring species richness provide detailed local data, it is hard to upscale this information to large geographical areas (Duro et al. 2007). Remote sensing tools (with NDVI playing an important role) offer opportunities for such large area descriptions of biodiversity in a systematic, repeatable, and spatially exhaustive manner (Duro et al. 2007; Turner et al. 2003).

NDVI plays an important role in the development of land-cover maps—an important tool in the *direct approach* or *first-order analysis* of species occurrence (Turner et al. 2003). Depending on the scale, biome, and ecosystem in question, land-cover maps provide implicit or explicit data on the composition, abundance, and distribution of individual or assemblages of species (Duro et al. 2007; Pettorelli et al. 2014; Turner et al. 2003). Data derived from vegetation productivity, in association with other environmental parameters (climatic and geophysical), are statistically related to species abundance or occurrence data (Duro et al. 2007; Pettorelli 2013). One example is the use of AVHRR-derived NDVI to explain the spatial variability of species richness of birds at a quarter degree spatial resolution in Kenya, finding a strong positive correlation between species richness and maximum average NDVI (Oindo et al. 2000). NDVI also contributes to the *indirect approach* to measuring species composition, abundance, and distribution. Different aspects of vegetation condition (derived from vegetation indices such as NDVI) contribute to the mapping of environmental variables which provide indications (through biological principles) of species composition, abundance, and distribution (Duro et al. 2007; Pettorelli et al. 2014). A high resource abundance (indicated by high NDVI values derived from NOAA/AVHRR satellite imagery) was used to explain the occurrence and distribution of the devastating locust specie *Schistocerca gregaria* in Mauritania (Despland et al. 2004).

3.9 Monitoring Ecosystem Resilience

The use of NDVI in vegetation monitoring and assessment is aimed at improving our understanding, predictions, and impacts of disturbances such as drought, fire, flood, and frost on global vegetation resources (Pettorelli et al. 2005, 2014). The use of the NDVI to monitor vegetation and plant responses to environmental changes at the level of trophic interactions constitutes one of the main uses of the NDVI in nature and conservation research. The application of the NDVI as a resilience indicator has been applied in numerous studies, such as reported by Díaz-Delgado et al. (2002), Simoniello et al. (2008), and Cui et al. (2013). Diaz-Delgado et al. (2002) used NDVI values derived from Landsat imagery to assess the recovery of Mediterranean plant communities after recurrent fire disturbances between the periods 1975 and 1993. They concluded (among other things) that the use of time-series

NDVI and other imagery products can be useful in understanding the resilience of Mediterranean plant communities and postfire vegetation dynamics over large regions and long time periods (Díaz-Delgado et al. 2002). Simoniello et al. (2008) characterized the resilience of Italian landscapes using NDVI trends to estimate mean recovery times of vegetation to different levels of anthropic pressure. They concluded that with 8-km AVHRR-NDVI data and remote sensing techniques, substantial details on vegetation cover activity (pointers to its resilience) at local scale could be captured, even in ecologically complex territories such as that of the Italian peninsula (Simoniello et al. 2008).

Cui et al. (2013) used Landsat TM and Landsat MSS time-series data to characterize land-cover status as a proxy for ecosystem resilience. They observed that the state of Southern African ecosystems and their response to a climatic shock (drought conditions) could be quantified in terms of vegetation amount and heterogeneity (Cui et al. 2013). Gibbes et al. (2014) used 28 years of AVHRR-MODIS NDVI time-series data in conjunction with global gridded monthly time series of modeled rainfall to determine the resilience of ecological systems in the Kavango–Kwandu–Zambezi catchments. Besides highlighting the explicit vegetation–precipitation linkages across this highly vulnerable region, the study underlined the important role played by precipitation in modulating conditions of the savanna ecosystems (Gibbes et al. 2014). Another application of NDVI to assess the resilience of ecosystems involves comparing stable-state NDVI trends to post-disturbance (from events such as fire, flooding, and hurricanes) NDVI trends to determine differences in ecosystem productivity across spatial-temporal scales (Renschler et al. 2010). These studies confirm the ability of remote sensing-derived NDVI, in combination with rainfall data, to detect land degradation processes that can be related to the resilience of ecosystems and landscapes.

Chapter 4
Limits to the Use of NDVI in Land Degradation Assessment

During the past half century, NDVI has been widely used for vegetation mapping and monitoring as well as in the assessment of land-cover and associated changes. This is because remotely sensed satellite-derived datasets provide spatially continuous data (data that are not sampled at individual points) and yield time-series signatures from which temporal patterns, trends, variations, and relationships may be derived (Jacquin et al. 2010). This has not prevented the misuse of NDVI—care is needed in the use of any scientific methodology.

As a spectral index of vegetation, NDVI provides the most direct quantification of the fraction of photosynthetically active radiation (fPAR) that is absorbed by vegetation (Running et al. 2004) (Figs 4.1 and 9.2). The convenience of satellite-derived NDVI and techniques of remote sensing for monitoring vegetation cover and assessing vegetation condition has been demonstrated at spatial scales from local to global and in diverse fields of environmental studies:

- Desertification (Olsson et al. 2005; Sternberg et al. 2011; Tucker and Nicholson 1999; Wessels et al. 2004; Symeonakis and Drake 2004)
- Drought assessment and monitoring (Anyamba and Tucker 2012; Bandyopadhyay and Saha 2014; Karnieli et al. 2010; Liu and Juárez 2001)
- El Nino impacts on ecosystems (Liu and Juárez 2001)
- Monitoring and assessment of regional to global changes in land cover and land use (Achard et al. 2007; Bradley and Mustard 2008; Cook and Pau 2013; Field et al. 1995; Lambin and Ehrlich 1997; Prince and Goward 1995; Rouse et al. 1974)
- Ecosystem health and services (Pettorelli et al. 2014; Zhang et al. 2013; Bai et al. 2013)

Nonetheless, the use of NDVI to discriminate directly between degraded and non-degraded areas can be challenging, both in implementation and interpretation. Wessels et al. (2004) studied non-degraded and degraded areas in northeastern South Africa, both exposed to identical rainfall regimes, paired and monitored over 16 growing seasons, and concluded that, in some cases, degraded areas were no less

© The Author(s) 2015
G.T. Yengoh et al., *Use of the Normalized Difference Vegetation Index (NDVI) to Assess Land Degradation at Multiple Scales*, SpringerBriefs in Environmental Science, DOI 10.1007/978-3-319-24112-8_4

stable than non-degraded areas. This is a call for caution in properly contextualizing land degradation assessments based on NDVI-, NPP-, or RUE-derived indices (see Sect. 5.5). It is our recommendation in this report that all NDVI studies use MODIS NDVI data as the benchmark. If identical NDVI trends between or among different NDVI datasets are not found, something is incorrect, and it is not the MODIS NDVI data. So all NDVI3g land surface studies be compared to MODIS NDVI data from the overlapping period (see Fig. 9.1). Failure to perform this inter-comparison can only lead to confusion.

Other issues also need to be considered in the use of NDVI for land degradation assessments:

- *The contentious issue of NDVI saturation at higher LAIs*: It has been argued that the NDVI signal from tropical evergreen forests is saturated so that there is a low signal/noise ratio. This is reported to occur when NDVI is related to LAI through a linear or exponential regression model (Schlerf et al. 2005). We are using NDVI as a surrogate for photosynthetic capacity. When photosynthetic capacity is at a maximum, there will be no change in NDVI because there is no change in photosynthetic capacity—because all visible photons have been absorbed. This is not saturation because primary production or photosynthesis is driven by light absorption. When there is no more light to be absorbed in high leaf density situations, photosynthesis is at a maximum and cannot increase (Fig. 4.1). Furthermore, integrated NDVI is directly related in a linear fashion to integrated fluorescence (Fig. 9.2). If NDVI saturation was real, the relationship in Fig. 9.2 would not occur.

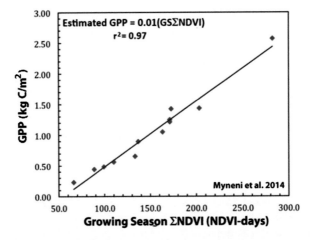

Fig. 4.1 Comparison between integrated gross primary production from 12 flux towers and integrated NDVI from MODIS Terra for the respective growing seasons where the flux towers were situated. This demonstrates the strong relationship between NDVI and primary production which is directly related to chlorophyll abundance and energy absorption (Myneni et al. 1995, 2014). There is no saturation of NDVI with respect to photosynthetic capacity (also see Fig. 9.2)

- Caution is needed in the use of vegetation indices to estimate LAI because there is no unique relationship between LAI and any particular vegetation index, but rather a set of relationships, each depending on the architecture of the plants in question (Haboudane et al. 2004). A LAI of a grass canopy cannot be compared to the same LAI of a broadleaf forest—the former has vertical leaves and the latter has horizontal leaves. NDVI is directly related to primary production and energy absorption (Myneni et al. 1995) and not to LAI.
- *It's hard to separate the effects of climate from the effects of land degradation:* Wessels et al. (2007) used the trend of residuals (RESTREND) to distinguish human-induced land degradation from the effects of rainfall variability. GLADA's empirical screening using RUE identified much the same patterns as RESTREND and has the advantage that it translates to NPP for economic analysis (see discussion in Sect. 5). Conijn et al. (2013) disentangled climate from other factors affecting NDVI by modeling biomass production independently according to crop characteristics and global data for climate, soils, and land use. Combining NDVI trends and modeled changes in biomass yields four scenarios: (a) positive $\sum$NDVI and positive biomass change (where greening might be explained by improving weather), (b) positive $\sum$NDVI and negative biomass (in spite of worsening weather, greenness has increased, thanks to management or atmospheric fertilization), (c) negative $\sum$NDVI and positive biomass (greenness declines against a trend of expected increase, so land degradation or land-use change has outweighed favorable weather), and (d) negative $\sum$NDVI and negative biomass (declining greenness may be explained by worsening weather) (Conijn et al. 2013). Whether the benefits of clearer separation of the climatic drivers is worth the substantial effort required is debatable, bearing in mind that the spatial variability of rainfall in drylands makes interpolation of the sparse point measurements problematic and the limitations of biomass modeling using a limited number of vegetation types and predefined management. It is no easier to verify changes in calculated biomass using independent data than to verify changes in NDVI.
- *Cloudiness:* The traditional approach of dealing with cloud cover has been to use Maximum Value Composites (MVC) (Holben 1986) which minimizes cloud contamination, reduces directional reflectance and off-nadir viewing effects, minimizes sun-angle and shadow effects, and minimizes aerosol and water-vapor effects. MVC requires that a series of original daily observations of multitemporal geo-referenced satellite data be processed into NDVI images pixel by pixel. Each NDVI value is inspected, and only the highest value is retained for each pixel location to eventually form part of an MVC image. MVC has been used in the production of the GIMMS NDVI 8-km dataset (Tucker et al. 2005). This compositing approach was necessary because the needed atmospheric variables were not present in the early part of this record for explicit corrections. To avoid time-series bias, the same processing approach has been applied to the entire 1981–2014 NDVI3g dataset.

This and any other compositing procedure may give rise to bias if a single false high is registered (Pettorelli et al. 2005). All compositing approaches may underestimate photosynthetic capacity under cloudy conditions, high aerosol situations, and the presence of snow cover. To eliminate these possibilities of bias, de Jong et al. (2011a) used the HANTS algorithm to remove residual cloud effects by applying Fourier analysis complemented by detection of outliers that were replaced by a filtered value. This approach identifies the high frequency "noise" components in NDVI time series data sets and removes them. All vegetation index data sets suffer from these problem and may be corrected using HANTS. Comparison of global NDVI trends using the HANTS-reconstructed data with the original GIMMS data showed no measurable difference (de Jong et al. 2011a).

– *Autocorrelation nullifies trend analysis:* Autocorrelation (whereby any individual value is influenced by the preceding values) is avoided by using annual $\sum$NDVI rather than the fortnightly GIMMS values, but this entails a loss of information; for example, we cannot analyze subtly changing seasonal responses of the NDVI signal that may indicate the nature of any degradation. De Jong et al. (2011a) applied the nonparametric Mann–Kendall model that is unaffected by autocorrelation to GIMMS NDVI data and normalized the data for seasonal variations in phenology rather than calendar years (which should be better in the southern hemisphere where growing seasons do not fall neatly within the calendar year). Linear regression measures annual accumulated photosynthetic activity, while Mann–Kendall measures the photosynthetic intensity of the growing season. Each has its own advantages, but the close similarity of the patterns of greening and browning revealed by the two models suggests that both are robust.

– *Direct assessment of some land degradation processes:* As discussed in Sect. 3.5, there are constraints in using NDVI to identify and map soil salinity: most of the salt remains below the soil surface so it cannot be detected on satellite images (Farifteh et al. 2006); surface salinity is very dynamic, and its detection can be blurred by vegetation and other surface features (Metternicht and Zinck 2003). The effects of sodicity and soil acidity are also impossible to distinguish from other limitations on the growth and productivity of vegetation.

– *Measurement of land degradation using NDVI trends underestimates the problem:* Farming everywhere is running down stocks of soil organic matter that supplies plant nutrients; maintains infiltration, available water capacity, and resilience against erosion; and fuels soil biodiversity. Over the last century, 60 % of soil and biomass carbon has been lost through land-use change. Chernozem, the best arable soils in the world, has lost 30–40 % of their organic carbon, yet they yield abundantly till a tipping point is reached and then the system flips—like the American Dust Bowl in the 1930s (Krupenikov et al. 2011). NDVI data do show that heavy use of fertilizer across much of China, the Indo-Gangetic Plain, Europe, the American mid-West, and southern Brazil is no longer accompanied by increasing production but may be concealing soil degradation.

Chapter 5
Key Issues in the Use of NDVI for Land Degradation Assessment

5.1 NDVI, NPP, and Land Degradation

A substantial body of research has established the correlation between NDVI and aboveground biomass, and knowledge of the theoretical basis for using satellite-derived NDVI as a general proxy for vegetation conditions has advanced (Mbow et al. 2014; Pettorelli et al. 2005; Sellers et al. 1994). Reduction of primary productivity is a reliable indicator of the decrease or destruction of the biological productivity, particularly in drylands (Wessels et al. 2004; Li et al. 2004). NPP expressed in g of C m^{-2} years^{-1} and quantifies net carbon fixed by vegetation. According to Cao et al. (2003), NPP is "the beginning of the carbon biogeochemical cycle," defined mathematically as in Eq. (5.1):

$$NPP = f(NDVI, PAR, fPAR, aPAR, LAI) \tag{5.1}$$

where fPAR is the fraction of absorbed photosynthetic active radiation, aPAR is the absorbed photosynthetic active radiation, and LAI is the leaf area index. Changes in NPP or, rather, its proxy NDVI induced by land degradation can be measured using a range of remote sensing techniques so remote sensing has become an essential tool for global, regional, and national studies of land degradation (Anyamba and Tucker 2012; Bai et al. 2008; Bajocco et al. 2012; de Jong et al. 2011b; Field et al. 1995; Horion et al. 2014; Le et al. 2014; Prince and Goward 1995). Many approaches have been developed to estimate NPP, notably the Global Production Efficiency Model (GLO-PEM) (Prince and Goward 1995), the Light-Use Efficiency (LUE) Model (Monteith and Moss 1977), the Production Efficiency Approach (Goetz et al. 1999; Goward and Huemmrich 1992), and the Sim-CYCLE (Ito and Oikawa 2002). And models have been developed to estimate NPP directly from remotely sensed NDVI at a global scale. Running et al. (2004) offered Eq. (5.2):

$$NPP = \Sigma \left(\varepsilon \times NDVI \times PAR - R_{lr} \right) - R_{g} - R_{m} \tag{5.2}$$

© The Author(s) 2015
G.T. Yengoh et al., *Use of the Normalized Difference Vegetation Index (NDVI) to Assess Land Degradation at Multiple Scales*, SpringerBriefs in Environmental Science, DOI 10.1007/978-3-319-24112-8_5

where ε is the conversion efficiency; PAR is photosynthetically active radiation; R_{lr} is 24-h maintenance respiration of leaves and fine roots; R_g is annual growth respiration required to construct leaves, fine roots, and new woody tissues; and R_m is the maintenance respiration of live cells in woody tissues. Drawing on this relationship, Bai et al. (2008) adopted an empirical relationship to translate NDVI trends to NPP trends for their proxy global assessment of land degradation (Eq. 5.3):

$$\text{NPP}_{\text{MOD17}}\left(kg\ C\ ha^{-1}\text{year}^{-1}\right) = 1106.37 \times \Sigma \text{NDVI} - 564.55 \qquad (5.3)$$

where $\text{NPP}_{\text{MOD17}}$ is the annual mean NPP derived from MODIS MOD17 Collection four data and sum NDVI is the 4-year (2000–2003) mean annual sum NDVI derived from GIMMS.

5.2   NDVI, RUE, and Land Degradation

The concept of rain-use efficiency (RUE), coined by Le Houerou (1984), is the ratio of aboveground NPP to annual precipitation. It tends to decrease with increase in aridity and potential evapotranspiration (Purkis and Klemas 2011; Symeonakis and Drake 2004; Le Houerou 1984). It has been observed that RUE is generally lower in degraded lands than in non-degraded lands (Symeonakis and Drake 2004); Fensholt et al. (2013) contend that RUE is "a conservative property of the vegetation cover in drylands, if the vegetation cover is not subject to non-precipitation related land degradation." Nonetheless, the use of RUE as an indicator for land degradation has been hotly contested on the grounds of methodology, differences in scale, and ecological contexts (Dardel et al. 2014; Fensholt et al. 2013; Wessels 2009; Fensholt et al. 2012; Wessels et al. 2007).

In the short term, vegetation reacts to natural rainfall variation so RUE needs to be examined over the long term to exclude false alarms (Nkonya et al. 2011). The common practice in estimating RUE is to use summed NDVI as an EO-based proxy for NPP (Fensholt et al. 2013), but the nature of the relationship between ΣNDVI and annual precipitation (proportionality, linearity, or nonlinearity) has been seen as an important consideration when estimating satellite-based RUE time series (Fensholt and Rasmussen 2011). In semiarid landscapes, where livestock farming is predominant, degradation from overgrazing often results in decreased or changes in the composition of vegetation communities and reduced rain-use efficiency (Diouf and Lambin 2001).

Using satellite-based ΣNDVI and annual precipitation, Fensholt and Rasmussen (2011) demonstrated that there is no proportionality, but sometimes a linear relation, between ΣNDVI and annual precipitation for most pixels in Sahel. The authors argue that this undermines the generalized use of satellite-based RUE time series as a means of detecting nonprecipitation-related land degradation.

RUE itself has been used as a proxy for land degradation (Safriel 2007; Symeonakis and Drake 2004) where RUE, itself, is not negatively correlated with rainfall. To stress the need for decoupling of precipitation and NDVI correlation in

RUE estimation and land degradation assessments, Fensholt et al. (2013) use the term *nonprecipitation-related land degradation*. This decoupling can be partly achieved by replacing annual ΣNDVI (a variable commonly and erroneously) used in RUE computations by a *small NDVI integral* that covers only the rainy season (not the whole year) and counting only the growth in NDVI in relation to some reference level. When this approach is applied to the African Sahel, Fensholt et al. (2013) find that positive RUE trends dominate most of the Sahel, which suggests that *nonprecipitation-related land degradation* is not widespread in the region.

While RUE can be used to normalize the effects of rainfall variability in the vegetation productivity signal when interpreting degradation trends (Landmann and Dubovyk 2014), the interpretation of RUE should be put in the proper environmental and land-use context; e.g., RUE is closely related to the scale of observation (Prince et al. 1998) and is not valid for land-use systems that show no rainfall–vegetation productivity correlations (Bradley and Mustard 2008). Given these caveats, the usefulness of RUE as a stand-alone indicator of vegetation productivity is limited. Another approach is through normalized cumulative RUE differences (CRD). The computation uses normalized monthly RUE with a Z-score normalization to correct for high outliers in the rainfall data (Landmann and Dubovyk 2014). Using 250-m MODIS NDVI data, this approach was employed to map vegetation productivity loss over eastern Africa between 2001 and 2011 (Landmann and Dubovyk 2014). The study concluded that 3.8 million ha of land experienced vegetation loss over the period, with an accuracy assessment of 68 % agreement between the rainfall-corrected MODIS productivity decline map and all reference pixels discernable from Google Earth and the Landsat-derived map and an accuracy of 76 % for deforestation. The study concluded that under high land-use intensities, the CRD showed a good potential to discern areas with *severe* vegetation productivity losses.

Dardel et al. (2014) used RUE residuals derived from linear regression as an indicator of ecosystem resilience in the Gourma region in Mali. This study made use of data from long-term field observations of herbaceous vegetation mass and GIMMS-3g NDVI data to estimate ANPP, RUE, and the ANPP residuals over the period 1984–2010. Counterintuitively, an increased runoff coefficient was observed over the same period of stable RUE. In Burkina Faso, an increase in discharge of rivers was first observed despite a reduction in rainfall—the *Sahelian Paradox* described in 1987 (Albergel 1988). Dardel et al. 2014 coined the term the *second Sahelian Paradox* to refer to the *divergence of these two indicators of ecosystem resilience (stable RUE) and land degradation (increasing runoff coefficient)* (Dardel et al. 2014).

5.3 Separating the Effects of Other Causes of NDVI Changes

Vegetative cover is a measurable indicator of ecosystem change, but the performance of vegetation depends on many micro- and macro-environmental factors (especially climate). Changes in vegetation reflect changes in both the natural factors that influence vegetation growth and performance, as well as human influences.

Hence, whereas NDVI can be a good indicator of NPP, separating the effects of climate variability on NDVI changes from those of land degradation is a challenge (Vogt et al. 2011). Different approaches, as described hereafter, have been used in recent studies.

One of the most popular techniques of application of NDVI in the assessment and monitoring of desertification is through the analysis of time-series NDVI images. In NDVI time-series analysis, linear models may be used to best fit the cyclic vegetation variation into a line (Eq. 5.4). The slope of the line can then be used to deduce the direction of vegetation variation (decrease or increase), as well as the strength of variation from the steepness of the trend line (e.g., no change, minimal, moderate, or severe change). In this case, the NDVI linear model will be

$$NDVI_t = a \cdot t + NDVI_0 \tag{5.4}$$

where a is the trend, $NDVI_0$ is constant, and t is time.

Evans and Geerken (2004) used linear regression on NDVI time series and rainfall to discriminate between the NDVI signal attributable to climatic conditions and that to human influence. The difference between the observed maximum NDVI and the regression-predicted maximum NDVI (referred to as residuals) was calculated pixel by pixel to identify the climate signal (the effect of precipitation) (Evans and Geerken 2004). Once the climate signal is identified and removed from the trends in vegetation activity, the remaining vegetation variations may be attributed to human activities. Positive trends in the vegetation represent areas of vegetation recovery, while negative trends constitute human-induced degradation of vegetation cover. RESTREND involves regressing ΣNDVI from annual precipitation and then calculating the residuals—the difference between observed ΣNDVI and ΣNDVI as predicted from precipitation (Fensholt et al. 2013; Fensholt et al. 2012; Wessels et al. 2007). RUE and RESTREND were compared using AVHRR NDVI from 1985 to 2003 and modeled NPP from 1981 to 2000 to estimate vegetation production in South Africa (Wessels et al. 2007). The study found that RUE was not a reliable indicator of land degradation. RESTREND was found to offer better prospects, but the study cautioned on the need for local-level investigations to identify the cause of negative trends. In a later study, Wessels et al. (2012) concluded that the RESTREND approach also has shortcomings, since the calculation of the residual NDVI is based on the assumption of a strong linearity between NDVI and rainfall over time, a relation which in the case of degradation during the period of analysis will be altered, thereby compromising the reliability of the RESTREND calculation.

Propastin et al. (2008) proposed the use of geographically weighted regression (GWR) between NDVI and precipitation for identifying the human-induced signal in NDVI time-series data. Using this method, the GWRs will describe the expected (predicted) NDVI for any particular climate signal. Deviations from the regression line by the observed NDVI would indicate vegetation changes that are attributable to stimuli other than climate (Propastin et al. 2008). Positive deviations indicate vegetation improvements, while negative deviations indicate declining in vegetation condition.

5.4   Abrupt Changes

Linear trends are easy to calculate, and in the early years of satellite NDVI, this was the only feasible approach. However, contrasting trends can balance out so it is important to ensure that the necessary assumptions for the determination of linear trends are met for each analysis. Citing De Beurs and Henebry (2005), Higginbottom and Symeonakis (2014) list them as (1) independence of the dependent variable, (2) normality in the model residuals, (3) consistency in residual variance over time, and (4) independence in residuals. Where there is need to separate NDVI time series into linear trend, seasonal components, and errors, nonlinear NDVI time-series models are used (Erian 2005). Key components of such nonlinear models (Eq. 5.5) would usually include a linear trend, stochastic component, external variables, periodic components (in terms of cycles and periodic trends), and white noise (Erian 2005, citing Udelhoven and Stellmes 2007).

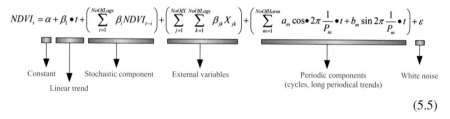

$$NDVI_t = \alpha + \beta_1 \bullet t + \left(\sum_{i=1}^{NoOfLags} \beta_i NDVI_{t-i} \right) + \left(\sum_{j=1}^{NoOfX} \sum_{k=1}^{NoOfLags} \beta_{jk} X_{jk} \right) + \left(\sum_{m=1}^{NoOfHarm} a_m \cos \bullet 2\pi \frac{1}{P_m} \bullet t + b_m \sin 2\pi \frac{1}{P_m} \bullet t \right) + \varepsilon$$

Constant — Linear trend — Stochastic component — External variables — Periodic components (cycles, long periodical trends) — White noise

(5.5)

The more than 30 years of NDVI record now available reveal many breaks of trend, even reversals. The Breaks for Additive Season and Trend (BFAST) approach has been developed to detect and capture these NDVI trend changes (de Jong et al. 2011c; Verbesselt et al. 2010a; Verbesselt et al. 2010b). The algorithm combines the decomposition of time series into seasonal, trend, and remainder component with methods for detecting changes. An additive decomposition model is used to iteratively fit a piecewise linear trend and a seasonal model (Haywood and Randall, 2008). According to De Jong et al. (2012), the general form of the model is Eq. (5.6):

$$Y_t = T_t + S_t + e_t : t \in T \qquad (5.6)$$

where, at time t (in the time series T), Y_t is the observed NDVI value, T_t is the trend component, and S_t is the seasonal component and the remainder component which contains the variation beyond what is explained by T_t and S_t. Using this method (Eq. 5.6), De Jong et al. (2011a) mapped the global distribution of greening to browning or vice versa; about 15 % of the globe witnessed significant trend shifts within the period studied. Common change detection methods average out this mixed trend effect, underestimating the trend significance (de Jong et al. 2011c).

Chapter 6
Development of Land Degradation Assessments

Early assessments of land degradation like the Global Assessment of Soil Degradation (GLASOD) (Oldeman et al. 1990) were compilations of expert opinion. They are unrepeatable and systematic data show them to be unreliable (Sonneveld and Dent 2009). Under the FAO/UNEP program *Land Degradation in Drylands (LADA)*, Bai et al. (2008) undertook a global assessment of land degradation and improvement (GLADA) by analysis of linear trends of climate-adjusted GIMMS NDVI data. GLADA, the first quantitative assessment of global land degradation, aimed to identify and delineate *hot spots of land degradation, and their counterpoint—bright spots of land improvement* (Bai et al. 2008). The study revealed that about 24 % of the global land area was affected by land degradation between 1981 and 2003. Humid areas accounted for 78 % of the global degraded land area, while arid and semiarid areas accounted for only 13 %. Cropland and rangelands accounted for 18 % and 43 %, respectively, of the 16 % of global land area where the NDVI increased. The authors observed a positive correlation between population density and NDVI but, also, a correlation between poverty and land degradation. They emphasized that NDVI cannot be other than a proxy for land degradation and that it reveals nothing about the kind of degradation or the drivers (Bai et al. 2008).

Potential false alarms caused by drought cycles and rising global temperatures were removed by screening the data using rain-use efficiency (RUE) and energy-use efficiency (EUE). RUE was estimated from the ratio of the annual sum NDVI to annual rainfall calculated from the VASClimO station-observed monthly rainfall data gridded to 0.5° latitude/longitude (Beck et al. 2005); EUE was represented by the ratio of NDVI and accumulated temperature calculated from the CRU dataset (Jones and Harris 2013; Mitchell and Jones 2005). The sequence of operations was:

1. Areas where biomass productivity depends on annual rainfall were identified as those with a significant positive relationship between NDVI and rainfall. In these areas, years of below-normal rainfall exhibit below-normal NDVI and also, usually, increased RUE. Where there is decreasing NDVI but steady or increasing

RUE, the loss of productivity was attributed to drought and these areas were masked. Where both NDVI and RUE declined, something else is happening and these areas were included in the next stage of analysis.

2. For the remaining areas where productivity is not limited by rainfall and, also, for those with a positive relationship between productivity and rainfall but declining RUE, greening and browning trends were calculated as *RUE-adjusted NDVI*. Similarly, EUE was used to separate trends caused by rising temperatures, the net result being a *climate-adjusted NDVI*.

3. Urban areas were masked (this makes little difference to the global results −0.5 % for the identified degrading land and 0.2 % for improving land). Irrigated areas were not masked; the separation of areas of positive and negative correlation with rainfall effectively separates wetlands, irrigated areas, and areas with surplus rainfall from the areas where unadjusted NDVI is a good measure of degradation and improvement. Humid areas have not been masked; unadjusted NDVI was used for all of those areas where RUE is *not* appropriate.

4. The T-test was used to test the significance of the linear regression; class boundaries were defined for 90, 95, and 99 % levels.

5. To arrive at a measure amenable to economic analysis, NDVI trend was translated into gain or loss of NPP by correlation with MODIS 8-day NPP data (Running et al. 2004) for the overlapping period (2000–2006).

6. Several indices of land degradation and improvement were compared with land cover, land use, and landform. Land-use change is a main driver of land degradation so it would be useful to undertake analysis of NDVI against change in land use and management, but there are no corresponding time series data for land use or land cover. GLC2000 (Bartholomé and Belward 2005) global land-cover and land-use systems of the world (FAO 2013) were used for preliminary comparison with NPP trends.

7. Soil and terrain: A global soil and terrain database at scale 1:1 million-scale was compiled using the 90 m-resolution SRTM digital elevation model and a dataset of key soil attributes for the LADA partner countries (ISRIC 2008a, b). Correlations between land degradation and soil and terrain were investigated in country studies.

8. Population, urban areas, and poverty indices: The CIESIN Global Rural–urban Mapping Project provides data for population and urban extent, gridded at 30 arc-second resolution (CEISIN 2004). Subnational rates of infant mortality and child underweight status and the gridded population for 2005 at 2.5 arc-minutes resolution (CEISIN 2007) were compared with indices of land degradation.

The picture revealed by GLADA was against received wisdom which reckoned that degradation was worst in the Sahel, the Amazon rain forest, and, more generally, in drylands. But the Sahel, Amazon, and drylands mainly showed increases in climate-adjusted NDVI. The areas hardest hit appeared to be Africa South of the Equator, Southeast Asia, the Pampas and Chaco regions in South America, North Central Australia, and swaths of the high-latitude forest belt extending across North America and Siberia. However, the identification of increases in the

Amazon by Bai et al. (2008) may be questioned in the light of more recent studies by Morton et al. (2014) showing that the apparent greening of Amazon forests revealed in optical remote sensing data is due to seasonal changes in NIR reflectance—an artifact of variations in sun-sensor geometry (Morton et al. 2014). The picture is different again when the same analysis is applied to the extended GIMMS3g dataset for 1981–2011; the differences are not just because of the longer run of data but because of changes in GIMMS data processing to correct better for the periodic replacement of AVHRR sensors (especially AVHRR 2 to AVHRR 3). Importantly, the processing of the latest GIMMS dataset does not assume stationarity (no overall change in NDVI) but, rather, reveals the underlying trends (Pinzon and Tucker 2014).

Chapter 7
Experts' Opinions on the Use of NDVI for Land Degradation Assessment

Methodological issues were raised by Wessels (2009) regarding the GLADA assessment, chiefly the interpretation of RUE outside arid and semiarid regions, growing season differences between the northern and southern hemisphere and their implications for calendar year summations of NPP, and issues of scale in the interpretation of AVHRR NDVI vs. MODIS NPP relationships. He also maintained that the RESTREND technique provided a more dependable alternative. In response, Dent et al. (2009) clarified that RUE was not being used as an indicator of land condition but simply to separate NDVI trends caused by drought in those areas where biomass potential is directly related to rainfall, essentially drylands. Regarding seasonal differences in growing season between the northern and southern hemispheres, there was no difference in the long-term trends when the hydrological year was used for the southern hemisphere. And, finally, the RESTREND approach was also applied to the GLADA data and showed no significant difference with the RUE-adjusted NDVI approach; the choice of the RUE-adjusted NDVI was made on account of its simplicity and amenability to economic evaluation (Dent et al. 2009).

7.1 NDVI: Rainfall Proportionality, an Important Consideration

Methodological weaknesses that might seem to question the applications of RUE in the Bai et al. (2008) study (as with many studies in this area) have to do with the lack of consideration given to the effect of lack of proportionality in the use and interpretation of RUE (computed from NDVI-derived NPP) as a proxy for identifying areas of land degradation. The theoretical basis for RUE assumes proportionality between NPP (as indicated by NDVI) and rainfall (Le Houerou 1984), meaning that a fixed ratio (RUE) exists despite changes in rainfall over time. NDVI and rainfall should intercept at zero to meet this assumption of proportionality with changes in the

© The Author(s) 2015
G.T. Yengoh et al., *Use of the Normalized Difference Vegetation Index (NDVI) to Assess Land Degradation at Multiple Scales*, SpringerBriefs in Environmental Science, DOI 10.1007/978-3-319-24112-8_7

rainfall (Dardel et al. 2014; Fensholt and Rasmussen 2011; Verón et al. 2005). The theoretical assumption of proportionality is important in understanding the functioning of the relationship between these two variables (rainfall and NDVI). Recent studies have shed light on the importance of considering proportionality in the use of RUE derived from relationship between NPP (derived from NDVI) and rainfall (Dardel et al. 2014; Fensholt and Rasmussen 2011). The practical application of the relationship between NDVI and rainfall does not lead to generally robust results characterized by proportionality between variables. One of the main reasons for this lack of robustness is that NDVI is never zero—NDVI is always slightly positive, even on bare soils (Verón et al. 2005; Fensholt and Rasmussen 2011). The assumption of a linear relationship between NDVI and rainfall will also not to be applicable in cases where vegetation growth requires a certain threshold of rainfall (Dardel et al. 2014), which is the case in many areas of the tropical savannah with a distinct rainy and dry season. The result in both cases (either on bare soils or in places where a threshold of rainfall is required for vegetation growth to be triggered) is that a linear relation between NDVI and rainfall might exist but no proportionality. This means that RUE is in fact not able to normalize vegetation productivity for varying rainfall and consequently RUE calculated in this case is sensitive to changes in rainfall over both space (since it will artificially trend to infinity values (Dardel et al. 2014)) and time (Fensholt and Rasmussen 2011; Fensholt et al. 2013). The result of the RUE dependency on rainfall in a time-series analysis will be that "*significant trends will emerge if rainfall undergoes temporal changes within this range of values*" (Dardel et al. 2014; Fensholt and Rasmussen 2011; Fensholt et al. 2013). If overlooking this fundamental inefficiency of RUE to normalize for rainfall changes in the case where proportionality between variables is absent, the direct use of RUE for trend analysis (or indirect use when masking out pixels due to a certain trend in RUE) may lead to misleading interpretations. So as not to include cases with such artificial trends simply reflecting a change in the rainfall regime rather than land degradation, Fensholt and Rasmussen (2011) restricted their analysis of trends in Sahelian rain-use efficiency to using only cases: (1) where no per-pixel temporal correlation between annual RUE and rainfall was found; and (2) where estimates of growing season ΣNDVI and annual rainfall correlation were statistically significant ($p < 0.05$).

7.2 Building on the GLADA Assessment

Recent studies have used different approaches to assess land degradation at different scales, some using the GLADA methodology at different scales. Bajocco et al. (2012) summed NDVI values on a pixel basis recorded for each year between 2000 and 2002 and computed the mean annual ΣNDVI as a surrogate for the total annual biomass production of the Mediterranean region. Le et al. (2014) used the long-term trend of interannual mean NDVI over the period 1982–2006 to delineate land degradation hotspots but cautioned that the use of proxies is subject to uncertainties which need to be understood and addressed. De Jong et al. (2011a) made use of the

GIMMS NDVI data to analyze global greening and browning, using three approaches: a linear model corrected for seasonality, a seasonal nonparametric model, and analyzing the time series according to vegetation development stages rather than calendar days. The trends found using the linear model approach corrected for seasonality were very close to those identified by Bai et al. (2008) applying a linear model to yearly mean values, but there was a substantial difference in results from the different models—a cautionary reminder of the importance of putting results within the context of the methods applied and of providing adequate metadata to aid interpretation and understanding of the results. de Jong et al. (2011a, b, c) also used the Harmonic Analysis of NDVI Time Series (HANTS) algorithm to remove residual cloud effects by applying Fourier analysis, complemented by detection of outliers that were replaced by a filtered value. Comparison of global NDVI trends using the HANTS-reconstructed data with the original GIMMS data shows no measurable difference—so GLADA is unaffected by cloud cover. Chinese researchers have made use of the GIMMS database and the GLADA methodology for several countrywide studies. In the process, some new indices were developed, such as the sensitivity index—the degree of reaction of NDVI to rainfall change in specific rainfall regions (see Annexes 3 and 4).

Nkonya et al. (2011) used the first difference econometric approach in studying the global extent of land degradation and its human dimensions. NDVI trends were used to represent land degradation or improvement, and the NDVI-derived global land cover change was overlain with poverty distribution to better understand the connection between land degradation and poverty. The study found consistencies with Bai et al. (2008) in the relationship between severe poverty and decrease in the NDVI in some but not all parts of Africa. In a rigorous analysis of greenness in semiarid areas, worldwide, using AVHRR GIMMS data from 1981 to 2007, Fensholt et al. (2012) found that semiarid areas are, on average, greening, but similar increases in greenness over the study period may have broadly different explanations and cautioned against general assertions of ongoing land degradation in semiarid regions.

Chapter 8
Main Global NDVI Datasets, Databases, and Software

Coarse spatial resolution datasets are invaluable at the global scale, but they lack the thematic and spatial detail required for habitat assessments at the country level and for finer-resolution assessments such as vegetation species distribution or high-quality forest-change monitoring. Mapping, monitoring, and assessments at the national and subnational level are performed using moderate-resolution sensors such as Landsat, ASTER, SPOT HRV, and IRS with spatial resolutions from 15 to 60 m. Newer, high-resolution optical sensors (5 m or better) provide enough spatial and spectral detail to discriminate between individual trees and, in some cases, species, but high-resolution imagery is prohibitively costly (see Annex 7) for many national governments and research institutions (Strittholt and Steininger 2007).

8.1 Main NDVI Datasets

Significant research effort has been invested in processing satellite sensor data into NDVI. The most common sensor used in these initiatives is the AVHRR sensor on board the NOAA satellites which enables assembly of global NDVI datasets. The development of these datasets by different research groups involves diverse schemes, protocols, and algorithms for corrections and processing (Scheftic et al. 2014). As a result, the environmental change community currently has a range of datasets that may be used for a variety of applications (Table 8.1).

The most widely used global NDVI datasets are the Global Inventory for Mapping and Modeling Studies (GIMMS); NOAA/NASA Pathfinder (PAL); the Long-Term Data Record (LTD); and the Fourier-Adjusted, Sensor and Solar zenith angle corrected, Interpolated, Reconstructed (FASIR) adjusted (NDVI) (see Annex 1).

The *Global Inventory for Mapping and Modeling Studies (GIMMS)* dataset is the most updated global time-series NDVI product (Fensholt and Proud 2012). It has a temporal resolution of 2 weeks (24 scenes/year) and a spatial resolution

© The Author(s) 2015
G.T. Yengoh et al., *Use of the Normalized Difference Vegetation Index (NDVI) to Assess Land Degradation at Multiple Scales*, SpringerBriefs in Environmental Science, DOI 10.1007/978-3-319-24112-8_8

Table 8.1 Commonly utilized normalized difference vegetation index (NDVI) datasets (modified from Higginbottom and Symeonakis 2014)

Name	Sensor	Time span	Time step	Resolution
Pathfinder (PAL)	AVHRR	1981–2001	10-day	8 km
Global Vegetation Index (GVI)	AVHRR	1981–2009	7-day	4 km
Land Long-Term Data Record (LTDR)	AVHRR	1981–2013	Daily	5 km
Fourier-Adjusted, Sensor and Solar zenith angle corrected, Interpolated, Reconstructed (FASIR)	AVHRR	1982–1998	10-day	0.125°
GIMMS	AVHRR	1981–2006	15-day	8 km
GIMMS3g	AVHRR	1981–2015	15-day	8 km
S10	SPOT-vegetation	1998+	10-day	1 km
EM10	ENVISAT-MERIS	2002–2012	10-day	1/1.2 km
SeaWiFS	SeaWiFS	1997–2010	Monthly	4 km
MOD (MYD)13 A1/A2	Terra (Aqua)	2000+	16-day	500 m/1 km
MOD13 (MYD) A3			Monthly	1 km
MOD13 (MYD) C1/C2			16-day/monthly	5.6 km
MOD13 (MYD) Q1	MODIS		16-day	250 m
MEDOKADS	AVHRR	1989+	Daily	1 km

of approximately 8 km. The GIMMS NDVI3g dataset (Pinzon and Tucker 2014) now comprises more than 33 years of data corrected for instrument calibration, variations in solar angle and view zenith angle, stratospheric aerosols from major volcanic eruptions, and other effects not related to vegetation change. Cloud and haze effects are minimized by taking the highest fortnightly value within composite 8 km blocks of pixels (Holben 1986).

The *NOAA/NASA Pathfinder (PAL) NDVI* dataset was created from the PAL 8 km daily product (Green and Hay 2002; James and Kalluri 1994). The PAL 8 km daily data were spatially re-sampled, based on maximum NDVI values from AVHRR Global Area Coverage data which have a minimal resolution of 4 km (De Beurs and Henebry 2005; James and Kalluri 1994). The PAL Global 10-day composite NDVI product is part of the Pathfinder Land dataset archived at the Goddard Earth Sciences, Distributed Active Archive Center (GES-DAAC) (Green and Hay 2002). The data have been corrected for changes in sensor calibration, ozone absorption, Rayleigh scattering, and sensor degradation after prelaunch calibration and has been normalized for changes in solar zenith angle (James and Kalluri 1994). The dataset is not continually being processed, and data after 1999 are not accessible online at the GES-DAAC website of the Goddard Space Flight Center.

Long-Term Data Record (LTD) is a global daily dataset of 0.05° (about 5 km ground spatial distance) developed by the NASA-funded LTDR Project. The dataset is currently at its fourth version and available for the period 1981–2013 from the reprocessing of the N07-N18 AVHRR data (Pedelty et al. 2007). The current version includes records from the processing of data from NOAA-16 and NOAA-17 lengthening the LTDR records from AVHRR to 2013.

The Fourier-Adjusted, Sensor and Solar zenith angle corrected, Interpolated, Reconstructed (FASIR) adjusted NDVI datasets are products of the International Satellite Land-Surface Climatology Project, Initiative II (ISLSCP II) data collection, developed to provide a 17-year satellite record of monthly changes in the photosynthetic activity of terrestrial vegetation for use in general circulation climate models and biogeochemical models (Sellers et al. 1994; Sietse 2010). The NDVI collections are provided in data files at spatial resolutions of 0.25, 0.5, and 1.0° latitude/longitude. FASIR adjustments concentrated on reducing NDVI variations arising from atmospheric, calibration, view, and illumination geometries and other effects not related to actual vegetation change (Sietse 2010).

The *Moderate-Resolution Imaging Spectrometer (MODIS)* is an extensive program using sensors on both the Terra and Aqua satellites, each of which provide complete daily coverage of the Earth. Started in January 2000, the MODIS sensor provides vegetation indices (NDVI and EVI) produced globally on 16-day intervals at three resolutions (250, 500, and 1000 m). The MODIS NDVI data are fully consistent spatially and temporally with AVHRR-NDVI products (Tucker et al. 2005). Comparisons between AVHRR and MODIS NDVI products over a wide range of vegetation types have shown a very high correlation, $r > 0.9$ according to (Gallo et al. 2005). See also Fig. 9.1. A complete collection of MODIS Land products can be accessed freely online either from USGS or NASA sites (see Annex 7).

Before deciding on the data to be used for NDVI-related analysis, users should reflect on a number of questions. According to Khorram et al. (2012), some of the most important of these questions include:

- What kind of remotely sensed data do I need? More specifically, what types of data are available from today's remote sensing instruments, and what are their strengths and limitations?
- How must my chosen data be prepared prior to analysis? What are the appropriate processing and/or analytical methods?
- What is the accuracy of the output products I have created? Is that accuracy sufficient for my ultimate objectives?

8.2 Quality-Related Considerations

The potential for using free data for assessment and monitoring of environmental change (principally forest cover change) at the global level has been most clearly demonstrated for Landsat products. The key challenges for creating global products on forest cover and cover change are the processes and tools for atmospheric correction, proper calibration coefficients, working with different phenologies between compilations, terrain correction, accuracy assessment, and the automation of land cover characterization and change detection (Townshend et al. 2012). Most of the commonly used datasets mentioned above (such as the PAL, GIMMS, LTDR, and the FASIR) have undergone many evaluations and intercomparisons on a range of criteria (Beck et al. 2011; Fensholt and Proud 2012).

Beck et al. (2011) undertook a global intercomparison of the four AVHRR-NDVI datasets (PAL, GIMMS, LTDR, FASIR) against Landsat imagery for the period 1982–1999, finding significant differences in trends for almost half of the total land surface. The PAL and the LTDR (Version 3) datasets lacked calibration; GIMMS had the best calibration and was the most accurate in terms of temporal change. In a study investigating whether vegetation trends derived from NDVI and phenological parameters are consistent across products, Yin et al. (2012) compared GIMMS and SPOT–VGT-derived NDVI. Strong similarities were found in interannual trends and, also, in trends of the seasonal amplitude and annual sum NDVI. But there were significant discrepancies between NDVI-derived trends based on phenological parameters such as amplitude (maximum increase in canopy photosynthetic activity above the baseline) and integral of NDVI (canopy photosynthetic activity across the entire growing season) (Yin et al. 2012). These correspond to seasonal vegetation cycles revealed by GIMMS and SPOT VGT. The study attributed these discrepancies to variables such as land cover and vegetation density. Such discrepancies highlight the need for appropriate and rigorous preprocessing when working with data from different remote sensing systems.

8.3 Precipitation Datasets

Various precipitation datasets are used in combination with NDVI data in many earth science applications. Among the most widely used of these datasets are the Modern-Era Retrospective Reanalysis for Research and Applications (MERRA), Interim Reanalysis (or ERA-Interim Reanalysis), Global Precipitation Climatology Project (GPCP), Africa Rainfall Climatology, and the VASClimO (Table 8.2).

The *Modern-Era Retrospective Reanalysis for Research and Applications (MERRA)* is a NASA reanalysis for the satellite era using the Goddard Earth Observing System Data Assimilation System Version 5 numerical weather and climate model. The Project focuses on historical analyses of the hydrological cycle on

Table 8.2 Commonly used precipitation datasets for earth science and environmental applications

Precipitation data	Time span and scale	Reference
NASA MERRA	1979–present at 0.5° × 0.5°	Rienecker et al. (2011)
ERA-Interim	1979–present at 80 km	Dee et al. (2011)
GPCP	1979–2012 at 1.0° × 1.0°	Huffman et al. (2009)
African Rainfall Climatology	1983–2013 at 0.1° × 0.1°	Novella and Thiaw (2013)
VASClimO	1951–2000 at 0.5°, 1.0°, and 2.5°	Beck et al. (2005), Schneider et al. (2008)
TRMM	1997–present at 0.25° × 0.25°	Gentemann et al. (2004)

a broad range of weather and climate time scales and places the NASA EOS suite of observations in a climate context. This dataset has a spatial resolution of $0.5° \times 0.5°$ from 1979 to the present (Rienecker et al. 2011).

The *Interim Reanalysis* (or ERA-Interim Reanalysis) output comes from the European Centre for Medium Range Weather Forecasts. It is a global atmospheric reanalysis from 1979, continuously updated in real time through the present with a spatial resolution of 80 km (Dee et al. 2011).

The *Global Precipitation Climatology Project (GPCP)* Version 2.2 is a blend of precipitation gauge data and satellite data taking advantage of the strengths of each data type. These data are $1° \times 1°$ and run from 1979 to 2012 (Huffman et al. 2009). Tropical Rainfall Measuring Mission or TRMM data form the basis of the GPCP dataset and are blended with station data to improve the rainfall accuracies.

The *Africa Rainfall Climatology* Version 2 is a gridded, daily 30-year (1983–2013) precipitation dataset at $0.1° \times 0.1°$ spatial resolution produced by NOAA's Climate Prediction Center (Novella and Thiaw 2013), produced using an operational rainfall estimation algorithm now updated to Rainfall Estimates Version 2 (Novella and Thiaw 2013).

The *VASClimO* is a global dataset of station-observed precipitation produced by gridding 9343 homogeneity-checked station time series of precipitation for the period 1951–2000 (Rudolf et al. 2005). It provides a globally gridded total monthly precipitation from January 1951 to December 2000 at three resolutions, $0.5° \times 0.5°$, $1.0° \times 1.0°$, and $2.5° \times 2.5°$, and is updated by the GPCC full-data reanalysis product version 4 (Schneider et al. 2008).

Tropical Rainfall Measuring Mission (TRMM) dataset is obtained by active and passive microwave measurements derived by instruments on board the Tropical Rainfall Measuring Mission's (TRMM) Microwave Imager (TMI). Besides measuring rain rates, the TMI can also measure sea surface temperature (SST), ocean surface wind speed, columnar water vapor, and cloud liquid water. TRMM is a joint program between NASA and the National Space Development Agency of Japan (Gentemann et al. 2004).

8.4   NDVI Software

Besides the data requirement, the creation of NDVI products requires software. There are many software products that can be used to create NDVI products and they can be categorized according to the tool sets that they offer, as well as the tasks that each of them can accomplish (Steiniger and Hunter 2013). The general demands for software for geospatial analysis include capabilities for data capture and representation, data visualization and exploration, data editing, data storage, integration of data from different sources, data queries to select a subset of the data, data analysis, creation of new data from existing or available input data, data transformation, and elements of cartographic representation. Not all software have full out of the box capabilities of accomplishing these different tasks. Many software depend on extensions, plug-ins, and application program interfaces (sets of routines, protocols, and tools which specify how software components should interact and are used when programming graphical user interface components) to accomplish some of these tasks (Steiniger and Hunter 2013). The creation of NDVI products demands additional capabilities for the modification, as well as spectral transformation of aerial and satellite image data. The appropriate geospatial software for working with remote sensing products and the creation of NDVI products therefore need capabilities for image radiometric and geometric correction, filtering, georeferencing and ortho-rectification, mosaiking, vectorization, and image object extraction (Steiniger and Hunter 2013; Jensen 2007; Mather and Koch 2011).

Geospatial software may be commercial or free products. There are many commercial software products with different strengths and levels of specialization in the delivery of remote sensing services and products (see Annex 8A for some of the commonly used desktop software). There are also free or open-source software which offer a range of possibilities for geospatial analyses, including capabilities for the creation of NDVI products (see Annex 8B). Over the last decade, there has been a notable increase in the availability of free and open-source software projects for geographic data collection, storage, analysis, and visualization (Steiniger and Hunter 2013). The benefits of open-source software are well known and include cost savings, vendor independence, and open standards (Steiniger and Hunter 2013). There have been several initiatives to compile and make inventories of existing geospatial software products.[1]

Besides the desktop applications presented in Annex 8, there are other platforms through which NDVI products can be created. Steiniger and Hunter (2013) point to the importance of Server GIS and WebSphere Process (WPS) Servers which host software that expose GIS and remote sensing functionality typically found in desktop geographic information systems or remote sensing software. Such functionality does not require direct interaction from a user via a user interface (Cepicky and Becchi 2007).

[1] Initiatives such as those by Michael de Smith and Paul Longley of University College London, and Mike Goodchild of University of California, Santa Barbara, provide information on the main activity for which each software product is designed, their license status (commercial or free), as well as links to their access and further information: http://www.spatialanalysisonline.com/software.html. The Wikipedia page that compares geographic information systems and remote sensing software in terms of license status, operating system, and other operational specifications can be found at: http://en.wikipedia.org/wiki/Comparison_of_geographic_information_systems_software.

Chapter 9
Country-Level Use of Satellite Products to Detect and Map Land Degradation Processes

For ecological studies and environmental change research, Pettorelli et al. (2005) distinguish two main groups of satellite products:

(a) Long-term NDVI datasets including the coarse-scale (8–16 km resolution) NOAA–AVHRR time series extending from 1981 to the present and the small-scale Landsat–TM dataset extending from 1982; the use of Landsat products for land-use and land-cover change has been growing because Landsat has a relatively fine resolution for land-use change studies and wave bands extending across the visible, near-infrared, shortwave infrared spectrum (Townshend et al. 2012).

(b) Finer-scale but short-term NDVI time-series datasets which include MODIS–TERRA (250–1000 m resolution) extending from 2000 to the present and the 1 km to 300 m resolution SPOT–VGT dataset extending from 1998 to the present. However, these data are not available free of charge (see Annex 7).

Our approach to assessment of land degradation using satellite data depends on observing changes in total seasonal photosynthesis or primary production through time at continental scales, with the ability to disaggregate to national- and district-level scales when required. This disaggregation is necessary because all actions to halt land degradation must be implemented at the national or subnational scale. NDVI data exist globally at 8 km-resolution since 1981 from the AVHRR and at 250 m-resolution from MODIS since 2000. We recommend that all 8-km NDVI3g analyses should be complimented by comparisons with MODIS NDVI 250-m data for their overlap periods (Fig. 9.1).

What is the possibility of using other sources of primary production data? We could also possibly use the MODIS-derived net primary production product (MOD17) (Running et al. 2004) and chlorophyll fluorescence from the Greenhouse Gases Observing Satellite (GOSAT), the SCanning Imaging Absorption SpectroMeter

© The Author(s) 2015
G.T. Yengoh et al., *Use of the Normalized Difference Vegetation Index (NDVI) to Assess Land Degradation at Multiple Scales*, SpringerBriefs in Environmental Science, DOI 10.1007/978-3-319-24112-8_9

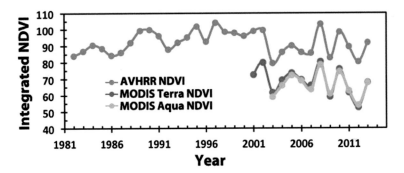

Fig. 9.1 Integrating NDVI values is directly related to gross primary production over the growing season for an area in Moldova. We have taken the NDVI values from Fig. 4.1 and numerically integrated them from the first of March to the end of October for 1981–2013 for the GIMMS NDVI3g dataset, for 2001–2013 for MODIS Terra NDVI, and for 2003–2013 for the MODIS Aqua NDVI data. Note very similar behavior in integrated NDVI values (NDVI days). There appears to be a break point between 2002 and 2003

for Atmospheric CHartographY (SCIAMACHY), or the Global Ozone Monitoring Experiment-2 (GOME-2) instruments (Joiner et al. 2012, 2013), which are alternative products to map and monitor land primary production. The MODIS NPP product is a global-modeled output product and, like many global products, performs less well when disaggregated to the national and district levels; its driving variables are not available at resolutions <1 km. The fluorescence products from SCIAMACHY, GOSAT, and GOME-2 satellites appear to be very useful for measuring primary production; SCIAMACHY data collection begun in early 2002 at a spatial resolution of 30×60 km (Gottwald et al. 2006); GOSAT data start in 2009 and are 10×10 km in spatial resolution (Joiner et al. 2011); GOME-2 data start in late 2006 and have a nadir spatial resolution of 0.5°×0.5° (Joiner et al. 2013).

The large spatial scale of these data is because fluorescence measurements are made within several Fraunhofer lines that are only 1 Angstrom (or 0.1-nm) wide so it is necessary to collect fluorescence data over large areas to get enough photons for an adequate signal-to-noise ratio. These coarse spatial scales make disaggregation to the subnational difficult. Recent studies by Joiner and a member of our team, Tucker (*submitted*), have shown that the time integral of fluorescence is linearly and very highly correlated to the NDVI time integral (Fig. 9.2). The NDVI advantage for land degradation studies is that land degradation can be studied over 33+ years with the GIMMS3g dataset at 8 km and for 15 years at 250 m from MODIS NDVI with the potential to downscale with NDVI data at 30 m from Landsat and at 1 m from commercial satellite data.

At present, there is insufficient time history of fluorescence to assess land degradation for these reasons: (1) Although SCIAMACHY fluorescence data started in 2002, their spatial resolution is 30×60 km which is very coarse resolution. (2) Fluorescence data from GOME-2 start in 2006 and from GOSAT in 2009 so we

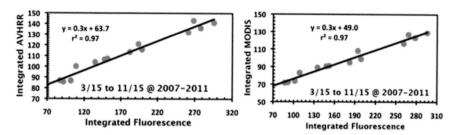

Fig. 9.2 Integrated AVHRR and MODIS NDVI compared to GOME-2 chlorophyll fluorescence for the Russian wheat growing areas of 51°–56°N×40°–54°E, 47–53°N×54°–60°E, and 50°–57°N×60°–72°E from March 15, 2007, to November 15, 2011. Three areas over 5 years provide a sample size of 15 (Yoshida et al. 2014). This figure contradicts the allegations that NDVI saturates and supports our use of the NDVI as being directly related to primary production

don't have enough time history to detect land degradation trends using these data. Twenty years on, satellite fluorescence data may be another tool for quantifying land degradation over large areas at a coarse scale, but, for the present, there is no alternative to NDVI in land degradation assessment.

Chapter 10
Challenges to the Use of NDVI in Land Degradation Assessments

Technological barriers: Currently, most global datasets useful for environmental applications are archived in databases that can be accessed using the Internet. These include the GIMMS, NOAA-PAL, LTD, and FASIR datasets. There are also free online data service platforms for executing preprocessing operations (such as data smoothing, spatial and temporal subsetting, mosaicking, and re-projection) of MODIS time-series vegetation indices (such as NDVI and EVI) on request. Currently, the Internet speed for many regions of the developing world remains too slow to enable effective access to these datasets or online processing.

Technical capacity: NDVI is a relatively simple index to compute and use in a number of environmental assessments (Liang 2005). However, when it comes to land degradation assessment, the use of NDVI can be problematic, both in implementation and interpretation (Wessels et al. 2004). Therefore, the analysts need to be properly equipped with the intellectual and technical skills to contextualize the problems of land degradation for particular cases and the interaction of key variables (NDVI-, NPP-, or RUE-derived indices) in the process.

Institutional and policy barriers: Effective use of satellite remote sensing products and technologies for a range of environmental assessments at the national level requires an appropriate institutional and policy framework. At the national level, this means the creation and effective management of a geo-information infrastructure that enables decentralization of information management through integration of geographic information and remote sensing systems. The level of integration could vary, depending on the setup of a country's administrative zones, the nature of land degradation being assessed, and the distribution of geo-information services in the country.

Barriers to effective knowledge management, decision support, and continuity: Given the complex array of environmental and socioeconomic processes involved in land degradation (Fig. 1.1), there is need for an effective system of knowledge and

© The Author(s) 2015
G.T. Yengoh et al., *Use of the Normalized Difference Vegetation Index
(NDVI) to Assess Land Degradation at Multiple Scales*, SpringerBriefs
in Environmental Science, DOI 10.1007/978-3-319-24112-8_10

information management. Information has to flow between and across sectors (such as agriculture, nature conservation, and other kinds of land use) for a proper interpretation of the distribution and trends of NDVI signals. A meaningful assessment of historic trends in land degradation or changes in land productivity requires continuity in the system of data collection, analysis, presentation, and activities related to each dataset or process. In many countries, frequent changes of government constitute a major constraint on the implementation of some programs and projects—ongoing policies, programs, and projects are often abandoned—creating a knowledge and continuity gap that may prove difficult to fill when these programs and projects are relaunched.

Economic and financial barriers: While the most popular NDVI datasets from major archives are free, effective access, processing, and use require some investment. The level of investments required depends on the scale of operations envisaged. While many governments in poor countries may lack the financial resources to put in place the full range of investments required for optimal access and use of existing NDVI databases, the costs of key investments in the sector are diminishing. This is especially true of hardware, some software, and Internet service costs. Investment in a professional and technical cadre is a bigger, longer-term issue, but home-grown expertise is essential if there is to be national ownership of the issue and the results of any assessment.

Chapter 11
Recommendations for Future Application of NDVI

11.1 In the Convention National Reporting

As discussed in the Introduction, both the UNCCD and the GEF use land cover to monitor land degradation. The UNCCD progress indicators (formerly known as impact indicators) should show progress made in achieving long-term benefits for people living in areas affected by desertification, land degradation, and drought, for affected ecosystems, and for the global environment. At its eleventh session, the COP adopted a refined set of six progress indicators (*Decision 22/COP.11; see Annex 6*) which will be used for the first time during the second leg of the fifth reporting process in 2016. Recommendations were made to the latest Conference of the Parties of the UNCCD (ICCD/COP(11)/CST/2) for refinements to the provisionally adopted set of impact indicators (Annex 6).

The findings of this report have implications for all three strategic objectives (SOs) of the UNCCD: SO-1 *to improve the living conditions of affected populations*, SO-2 *to improve the conditions of affected ecosystems*, and SO-3 *to generate global benefits through effective implementation of the UNCCD* (Table 11.1). Monitoring of drought using NDVI and NDWI could have implications for trends in access to safe drinking water (SO-1). It has been clearly shown that NDVI is a reliable measure of photosynthetic capacity and thus for monitoring trends in land cover and productivity of the land (SO-2). NDVI can also support reporting on global benefits related to trends in carbon stocks and biodiversity (SO-3), as shown in other sections of this report (also see Annex 2). Ideally, reporting on these indicators should be harmonized with reporting to the UNFCCC on carbon stocks and to the CBD on biodiversity indicators.

© The Author(s) 2015
G.T. Yengoh et al., *Use of the Normalized Difference Vegetation Index
(NDVI) to Assess Land Degradation at Multiple Scales*, SpringerBriefs
in Environmental Science, DOI 10.1007/978-3-319-24112-8_11

Table 11.1 UNCCD core indicators for national reporting

Indicator	Potential use of NDVI
Strategic objective 1: to improve the living conditions of affected populations	
SO-1. (1): Trends in population living below the relative poverty line and/or income inequality in affected areas	*Not applicable*
SO-1. (2): Trends in access to safe drinking water in affected areas	NDVI could be combined with the normalized difference water index (NDWI) to monitor drought and be linked to water use of land-use systems (see Annex 1)
Strategic objective 2: to improve the condition of affected ecosystems	
SO-2. (1): Trends in land cover	NDVI is the best tested vegetation index with the longest time series for monitoring of land-cover trends (33 years), which compensates for the low resolution. However, care needs to be exercised in interpretation of the results and the drivers of change (Annex 2)
SO-2. (2): Trends in land productivity or functioning of the land	The relationship between NDVI and biomass productivity has been well established in the literature. NDVI can be used to estimate land productivity and monitor such productivity over time (Annex 2)
Strategic objective 3: to generate global benefits through effective implementation of the UNCCD	
SO-3. (1): Trends in carbon stocks above- and belowground	NDVI can be used together with higher-resolution data to estimate trends in carbon stocks, e.g., REDD and SOC assessments (Annex 1)
SO-3. (2): Trends in abundance and distribution of selected species	NDVI can be used to monitor habitat fragmentation and connectivity which crucially affect the abundance and distribution of species (Annex 1)

11.2   In a Revised GEF Resource Allocation Methodology

Land cover is used as an indicator for all three GEF focal areas affected by the System for Transparent Allocation of Resources (STAR) that calculates country-specific allocations from each focal area[1]:

Land Degradation—the latest Global Benefit Index (GBI) for the land degradation (LD) focal area was designed to take into account three key factors in accordance with GEF mandate for financing: (1) the need to control and prevent land degradation in land-based production systems; (2) the challenge of combating desertification in the drylands, including the need for adaptation to drought risks; and (3) the need to address livelihoods of vulnerable populations. Proxy indicators were derived for each of these factors based on available data.

---

[1] GEF/POLICY: PL/RA/01, March 14 2013: System for Transparent Allocation of Resources (STAR).

With regard to factor (1), a quantitative estimate of land area (in km^2 or as percent of territory) affected by LD was used as a proxy indicator for *loss of ecosystem function and productivity*. The indicator was derived by Bai et al. (2008) using NDVI. Each country's share of the global total area affected was calculated for use in the GBI. The three indices were assigned weights as follows: 60 % to dryland area, 20 % to rural population, and 20 % to land area affected.

GBILD = (0.2 × global share of land area affected) + (0.6 × proportion of dryland area) + (0.2 × proportion of rural population)

Climate Change: For its land-use, land-use change, and forestry (LULUCF) component, it uses forest cover in hectares and absolute change in forest cover, as reported by countries to FAO. NDVI could potentially be used to strengthen this index as NDVI is strongly correlated with vegetation dynamics in humid areas.

Biodiversity: This index uses distribution of terrestrial eco-regions, including threatened eco-regions as monitored by WWF. Also here, the use of NDVI could improve data quality if it is used consistently.

Trends in NDVI could thus become an important part of a land-cover indicator cutting across three GEF focal areas and used as a proxy for productivity, carbon stocks, and biodiversity. With regard to the land degradation focal area, a revised GEF STAR should be based on all the six core indicators identified for the UNCCD Strategic Objectives (see Table 11.1 and Annex 6). However, with a more robust application of NDVI based on recent advances, this index might be given a greater weight in a revised STAR, as it can contribute to monitoring of five of the UNCCD indicators if applied consistently and using the most reliable datasets.

Chapter 12
Conclusion

This report examines the scientific basis for the use of remotely sensed data, particularly NDVI, in land degradation assessments at different scales and for a range of applications. It draws on evidence from a wide range of investigations, primarily from the scientific peer-reviewed literature but also non-journal sources.

Research in land degradation currently makes use of a wide variety of datasets of different geographical scales and spatial, spectral, and temporal resolutions. The availability of free data of continuous observations from medium to coarse spatial resolution satellite sensors continues to support a range of ecosystem models and environmental applications. At the global level, a few of these datasets stand out. In the context of NDVI-based potential for land degradation assessment, the AVHRR-derived GIMMS dataset is the most widely used product. In the short to medium term, the quality control required to make this dataset a transparent source for a range of environmental applications is guaranteed. In the same light, continuous updates to the archive to extend it well beyond 33 years will enhance the potential for this data to be used to identify longer-term trends and trend components.

The GLADA approach, which was based on an earlier version of GIMMS, has been widely adopted. Several studies have used the same and later versions of the GIMMS dataset, with or without the GLADA approach, to investigate an array of environmental issues. Many caveats raised initially, flagged by GLADA itself, have been dealt with. As new methods of data analysis are developed, and computers become more efficient in processing information, more questions that draw on the relationship between NDVI, RUE, EUE, and NPP may be explored. These questions could address growing and emerging concerns about the resilience of ecosystems, and the coupling of socio-ecological systems, as well as new horizons in environmental assessment and management. The GLADA approach and NDVI data archives offer the potential for assessment of the performance of different policy options and can inform the implementation of the UNCCD and the allocation of resources from its financial mechanism, the GEF.

© The Author(s) 2015
G.T. Yengoh et al., *Use of the Normalized Difference Vegetation Index (NDVI) to Assess Land Degradation at Multiple Scales*, SpringerBriefs in Environmental Science, DOI 10.1007/978-3-319-24112-8_12

As a tool, NDVI and related indices, as well as the GLADA approach, still have limitations. Beyond some of the technical weaknesses associated with implementation and interpretation, there are barriers to their effective use for national assessments. We note that, over recent years, hardware components as well as some software to support the use of NDVI in national assessments have become more accessible. Notwithstanding the fall in costs of hardware and software, there is need for national services to be staffed by personnel with the appropriate technical expertise. This is necessary for many reasons, including the ability to ask the right questions and use the appropriate tools and depth of analysis in answering them and the ability to produce end products that meet international standards for cross-country comparisons.

NDVI continues to be valid for measuring and reporting some of the key strategic objectives of the UNCCD and has the appropriate qualities for use as an indicator for a number of indices.

Appendix A
Inventory of Some Global and Sub-global Remote Sensing-Based Land Degradation Assessments

Scale	Land degradation domain	Time range	Data	Main findings
Senegal	Land-cover change and human well-being	1982–2008	GIMMS	Interpreting satellite-derived greening as an improvement of environmental conditions that may translate into more stable livelihoods and greater well-being of populations in the area may not always be justified (Herrmann et al. 2014)
Mali	Ecosystem resilience in relation to ANPP, RUE, and ANPP	1984–2010	GIMMS-3g	There is a divergence of two key indicators of ecosystem resilience: a stable RUE and increasing run-off coefficient, condition termed "the second Sahelian paradox" (Dardel et al. 2014)
Amazon	Land-cover change and global environmental implications	2000–2012	MODIS	The Amazon forest has declined across an area of 5.4 million km^2 since 2000 as a result of reductions in rainfall. If drying continues in this region, global climate change may be accelerated through associated feedbacks in carbon and hydrological cycles (Hilker et al. 2014)

(continued)

© The Author(s) 2015
G.T. Yengoh et al., *Use of the Normalized Difference Vegetation Index (NDVI) to Assess Land Degradation at Multiple Scales*, SpringerBriefs in Environmental Science, DOI 10.1007/978-3-319-24112-8

(continued)

Scale	Land degradation domain	Time range	Data	Main findings
Italy	Land-use and land-cover change	1984–2010	Landsat TM	Total regional forest cover increased by 19.7 %, consistent with National Forest Inventory data. Considerable forest expansion also occurred on degraded soils in drought-prone Mediterranean areas (Mancino et al. 2014)
Senegal	Land-use, land-cover change and environmental conditions	1982–2008	AVHRR	The interpretation of satellite-derived greening trend as an improvement or recovery is not always justified. For instance, the composition of the vegetation cover may show impoverishment even in the greening areas (Herrmann and Tappan 2013)
China	Soil organic carbon and salinization	2011	Landsat TM	Significant decrease in soil organic C and total N contents were observed with increasing salinity. Soil organic C and total N contents had significant positive correlations with the NDVI (Pan et al. 2013)
World	Trends and drivers of greenness in semiarid areas	1981–2007	GIMMS-g	Current generalizations claiming that land degradation is ongoing in semiarid areas worldwide are not supported by the satellite-based analysis of vegetation greenness (Fensholt et al. 2012)
West Africa	Soil erosion and land productivity	1982–2003	GIMMS	Multipronged assessment strategies offers better insights into different processes involved in land degradation (Le et al. 2012)
World	Vegetation greening and browning trends	1981–2006	GIMMS	Models confirm prominent regional greening trends identified by previous studies (de Jong et al. 2011a, b, c)

(continued)

(continued)

Scale	Land degradation domain	Time range	Data	Main findings
South Africa	Biodiversity monitoring and conservation	1995–2006	GIMMS	Change in productivity driven by rainfall as well as that caused by elephant populations has ramifications for biodiversity and also impacts on biodiversity (Hayward and Zawadzka 2010)
USA	Biodiversity monitoring and conservation	2005	MODIS	There is a significant positive correlation between species compositional dissimilarity matrices and NDVI distance matrices. Remotely sensed NDVI can be a viable tool for monitoring species compositional changes at regional scales (He et al. 2009)
China	Desertification and land surface conditions	1980, 1990, and 2000	Landsat MSS and TM/ETM+	Human activities might explain the expansion of desertification from 1980 to 1990. Conservation activities were the main driving factor that induced the reversion of desertification from 1990 to 2000 (Xu et al. 2009)
Zimbabwe	Land-use, land-cover change and degradation	2000–2005	MODIS	About 16 % of the country was at its potential production. Total loss in productivity due to land degradation stood at about 13 % of the entire national potential. Most of the degradation was caused by human land use, concentrated in the heavily utilized, communal areas (Prince et al. 2009)
Sahel	Desertification and drought—changing trends	1982–1999	PAL	A consistent trend of increasing vegetation greenness may be attributed to increasing rainfall, but also to factors such as land-use change and migration (Olsson et al. 2005)

(continued)

(continued)

Scale	Land degradation domain	Time range	Data	Main findings
Sahel	Drought and vegetation dynamics	1981–2003	AVHRR	The current trends of recovery in the Sahel are still far below the wetter conditions that prevailed in the region from 1930 to 1965. Current trend patterns therefore only reflect a gradual recovery from extreme drought conditions that peaked during the 1983–1985 period (Anyamba and Tucker 2005)
South Africa	Ecosystem resilience and stability of landscapes	1985–2003	AVHRR	While degraded areas were no less stable or resilient than non-degraded area, the productivity of degraded areas, per unit rainfall, was consistently lower than non-degraded areas. Degradation impacts tend to be reflected as reductions in productivity that varies along a scale from slight to severe (Wessels et al. 2004)
Global	Vegetation growth and NPP	1982–1999	PAL/GIMMS	Global changes in climate have reduced several critical climatic constraints to plant growth, leading to a 6 % increase in global NPP (Nemani et al. 2003)
China	Vegetation dynamics and variations in NPP	1981–2000	AVHRR	Increase in NPP of about 0.3 % per year and decrease in net ecosystem productivity between the 1980s and 1990s due to global warming (Cao et al. 2003)
Spain	Vegetation burning and recovery	1994	Landsat TM and MSS	There are different patterns of postfire recovery based on dominant plant species, severity of burn, and a combination of both factors (Díaz-Delgado et al. 2003)

Appendix B
Use of Remote Sensing-Derived Land Productive Capacity Dynamics for the New World Atlas of Desertification (WAD)

Courtesy of Michael Cherlet, Hrvoje Kutnjak, Marek Smid, and Stefan Sommer of the European Commission, Joint Research Centre, Ispra, Italy; and Eva Ivits, Eva of the European Environment Agency, Copenhagen, Denmark

Background and Rationale

The European Commission's Joint Research Centre of the (JRC), together with UNEP and supported by a global network of international research institutions and experts, is developing the new World Atlas of Desertification (WAD).

Monitoring and assessing land degradation dynamics involves extracting the most relevant information from time series of global satellite observations. The dynamics of the Earth's standing vegetation biomass is considered a valid approximation of land system productive capacity dynamics thus, also reflecting the underlying ecological conditions and possible constraints for primary productivity, such as soil fertility, water availability, land use/management, etc., and hence related to land degradation. In fact, reduction or loss of land productive capacity, mostly biological and/or economical, is one common denominator in the various definitions of land degradation.

The longest available satellite observation datasets with global coverage at 1-km resolution, from, e.g., the SPOT VGT sensor, have a continuous frequent temporal sampling over a long enough period, now 15 years, to allow extraction of proxy information on the phenology and seasonal productivity for each 1-km^2 area on Earth. Even longer continuous time series, more than 30 years, are available through the GIMMS NDVI product, dating back as far as to 1981. However, its spatial resolution is only 8×8 km, which may well be suitable for the analysis of broader land-atmosphere interaction but which has limitations for monitoring and assessing the human-ecosystem interactions at landscape level. These operate and function typically at smaller scales than can be depicted at the spatial resolution of GIMMS. Nevertheless, the length of this NDVI time series raises interest in ways to combine it with higher-resolution products for enhanced analysis.

© The Author(s) 2015   67
G.T. Yengoh et al., *Use of the Normalized Difference Vegetation Index
(NDVI) to Assess Land Degradation at Multiple Scales*, SpringerBriefs
in Environmental Science, DOI 10.1007/978-3-319-24112-8

Methodology for Time Series Processing and Analyses

Building on numerous studies that use time series of remotely sensed vegetation indices (e.g., NDVI, Fapar) as base layer, we expand this set of variables by calculating phenological metrics from time series of the vegetation index. Disaggegation of the original time series into phenological metrics yields additional information on various aspects of vegetation/land-cover functional composition in relation to dynamics of ecosystem functioning and land use (Ivits et al. 2012a). This can provide a quantitative basis to monitor such information on ecosystem dynamic equilibrium and change, envisaged to provide users with an independent measure on how ecosystems respond to external impacts, be it human induced or natural variability (Ivits et al. 2012b).

The resulting remote sensing-derived spatial layers are then combined with ancillary biophysical and socioeconomic information in order to flag areas that show signals of actual land degradation. This includes attributions to different levels of intensity and probability of major causes, which will include major land degradation/desertification issues and the associated land-use transitions considered in the WAD. They are summarized below (Sommer et al. 2011):

1. Overuse of agricultural land, intensification, inappropriate agricultural practices/non-SLM, and increased soil erosion
2. Increase in intensive irrigation, overuse of water resources, and salinization
3. Grazing mismanagement, overgrazing, and decreasing NPP in rangelands, soil degradation, and sand encroachment
4. Deforestation
5. Increased aridity or drought
6. Socioeconomic issues, changes in population distribution and density, rural migration/land abandonment, and urban sprawl
7. Uncontrolled expansion of mining and industrial activities, extensive air and water pollution, and soil contamination

Analysis of long-term changes and current efficiency levels of vegetative or standing biomass are combined into land-productivity dynamics according to Fig. B.1. According to this scheme, the evaluation proceeds as follows:

Analysis of long-term changes and current efficiency levels of vegetative standing biomass are combined into land productive capacity dynamics. Output from both the long-term change maps and current status map was combined with start levels at the beginning of the time series, with the state change of productivity, and with a relative productivity map based on the principles of local net scaling approach (Prince et al. 2009), relating all pixels within an ecosystem functional unit (Ivits et al. 2012a) to the productivity of the best performing samples of that respective unit.

This processing chain has been applied at global levels and results in five classes indicating areas of negative change, positive change or stability of land productive capacity dynamics (see Fig. B.2). The classes are interpreted as indicators of change

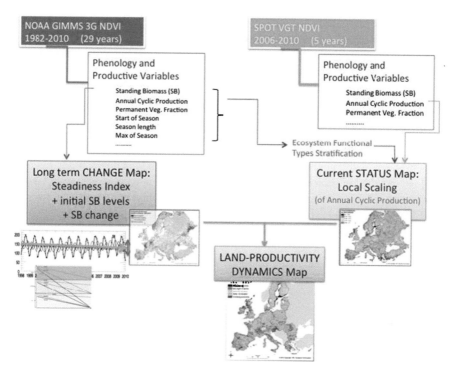

Fig. B.1 Processing scheme for deriving land productive capacity dynamics from the remote sensing time series (note that the approach was applied both to NOAA GIMMS 3G NDVI (Cherlet et al. 2013) as indicated above and also to 15 years SPOT VEGETATION NDVI 1999)

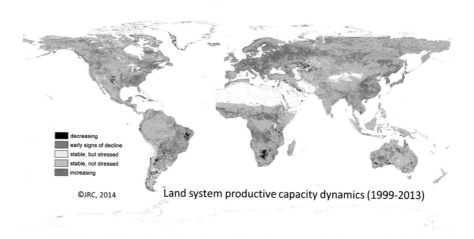

Fig. B.2 Global map of land system productive capacity dynamics derived from SPOT VEGETATION 15-year time series (1999–2013)

or stability of the land's apparent capacity to sustain its dynamic equilibrium of primary productivity during the given observation period, which is now further analyzed in relation to available information on land-cover/land-use and environmental change relevant to the issues listed in 1–7 above.

Preliminary Findings and Conclusions

This product should not be seen as a direct map of global land degradation, but as a globally mapped indicator which should be further evaluated in an integrated interpretation framework as proposed by the WAD or possibly also as outlined by the UNCCD ad hoc expert group (AGTE) for the new set of progress indicators.

Therefore, the land system productive capacity data are now further analyzed and evaluated in relation to available information on land-cover/land-use and environmental change relevant to the issues listed in Table 1.

An example of potential agricultural overexploitation of land is given below for Nigeria in Fig. B.3. Convergence of evidence is elaborated for interpreting the land productive capacity in the light of identifying and mapping ongoing critical land-use system transformations. Areas where the dynamics are decreasing are mostly areas where a number of land stress factors are coinciding, potentially threatening to a sustained use of the land. These are highlighted for further analysis. Stress

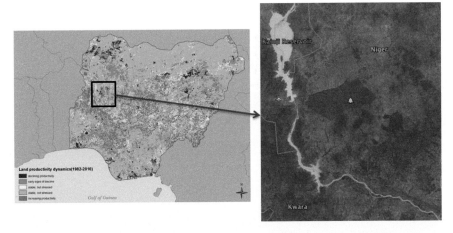

Fig. B.3 Interpretation example in Nigeria: The Kainji reservoir in Nigeria enabled expansion of agriculture in areas around the Dagida Game reserve (green area on the above map frame indicating a long-term stable land productive capacity). The map suggests that the capacity of the land to sustain a stable land-productivity dynamic equilibrium is declining in those areas of irrigated agriculture expansion around the protected areas; this highlights for further analysis to identify coinciding intensification and stresses factors can be natural, e.g., drought, or human induced such as overextraction of soil nutrients by demanding irrigated crops as shown below.

A broader statistical analysis and evaluation is also underway but needs to be carefully framed and integrated with the data and information provided within the thematic WAD chapters.

Nevertheless, some general observations of observed trends and patterns, here from the 15-year time series, appear coherent with other studies.

While excluding land areas with no significant vegetal primary productivity, i.e., hyperarid, Arctic, and very-high altitude mountain regions from statistics, it is evident that indications of decreasing land system productive capacity can be observed globally. About 20 % of the land surface, involving all vegetation cover types, are showing signs of decreasing land system productive capacity. Only 19.5 % of the considered vegetal productive land surface is cropland, of which 18.5 % show clearly decreasing trends or early signs, while for the rest of the non-cropland, 19.8 % is affected (which however accounts for 80 % of the overall area with declining land system productive capacity). Considering that strong efforts and resources are committed to maintain the productivity of agricultural land and the fact that there are clear limitations to further expansion of croplands, these figures are an issue of concern. The huge seminatural and rangeland areas affected (approx. 18 million sq. km), however, highlight the enormous dimension of the critical dynamically changing ecological conditions worldwide.

The picture is much more complex when breaking down the statistics to continental, regional, and subregional levels. As mentioned before, when entering into this necessary exercise, the increased use of additional thematic information for setting up a more stratified analytical approach is strongly recommended, although it is much-more time-consuming and cumbersome.

Nevertheless, when just looking at croplands at continental level, we notice substantial differences in the dimension/extension of potentially critical areas, which will require careful consideration and more in-depth analysis.

In Africa, about 20.6 % of the considered land surface is cropped of which 21.7 % show signs of decreasing land productive capacity. The relative proportion of concerned cropland is similar to that of seminatural rangelands. This is above the global average but not extremely. In turn, surprisingly, South America with about 21.4 % cropland shows a much higher percentage of potentially affected croplands, up to 31.9 %, also proportionally higher than for seminatural/rangelands areas (25.84 %).

Europe is the continent with the relatively highest extension of croplands, i.e., about 32 %, of which 15.75 % may be confronted with critical developments of land productive capacity, especially in the south of Eastern Europe but also the Iberian Peninsula, which is proportionally higher than for other land-cover/land-use types (10.3 %).

While human factors will need to be further analyzed at regional to subregional level, first analyses at global level dealt with the correlation of some spots of decreasing productive capacity in the shorter time-series product (i.e., 15 years, from 1999 to 2013) against actual global drought monitoring data. These revealed strong correlations of areas in southern Africa (between Botswana, Namibia and South Africa), northeastern Brazil, and Australia–Oceania with recent recurrent droughts. In this respect, it will be also a clear issue in the WAD to analyze in more in-depth the possible differences of dynamics between drylands and non-drylands.

Issues of uncertainty of remote sensing time-series-derived products as function of length of the time-series, spatial, and spectral characteristics will also need to be better addressed. It could be recommended that scientific groups developing monitoring products should join forces, for example, by addressing these issues in a kind of ensemble analysis aiming at ways to take benefit from combinations of longer time series and higher spatial detail, thus stimulating additional options and criteria for generating new elements of convergence of evidence.

Appendix C
Developments with GLADA

Courtesy of Zhanguo Bai, ISRIC – World Soil Information, Wageningen, The Netherlands

A Different View of the World

The original Global Assesment of Land Degradation (GLADA, Bai et al. 2008) set out to answer questions about the global extent, severity and cost of land degradation using GIMMS data for 1981–2003. In particular: Is land degradation a global issue or just a collection of local problems? Which places are hardest hit? Is it mainly a problem of drylands? Is it mainly associated with farming – or poverty? The results were against received wisdom and, therefore, contested. A somewhat different picture is revealed by further analysis of the extended GIMMS3g dataset for 1981–2011 (Fig. C.1). The differences are not just because of the longer run of data but because of changes in GIMMS data processing to better correct for the periodic replacement of AVHRR sensors as one satellite replaced another, especially AVHRR 2 to AVHRR 3 (Pinzon and Tucker 2014a, b). The original calibration of the data from successive AVHRR instruments assumed stationarity, i.e., that there was no underlying trend. We now know that NDVI exhibits complex long-term trends. Reprocessing of the whole dataset has removed biases introduced by the initial calibration and better reveals the underlying trends. Changes like this in the fundamental data do nothing for credibility, but we are confident that the fundamental data are now much improved – other datasets do not have the advantage of these corrections.

We can now give straight answers to our original policy questions:

– *Land degradation is a global issue* with 22 % of the land degrading over the last 30 years, representing a loss of net primary production of some 150 million tons of carbon but a loss of soil organic carbon orders of magnitude more.
– *The areas hardest hit* are Africa, especially south of the equator with an arm of degradation extending north to the Ethiopian highlands and two outliers in the Sahel—the Nile provinces of Sudan and Koulikoro Province, Mali; the Gran

© The Author(s) 2015
G.T. Yengoh et al., *Use of the Normalized Difference Vegetation Index (NDVI) to Assess Land Degradation at Multiple Scales*, SpringerBriefs in Environmental Science, DOI 10.1007/978-3-319-24112-8

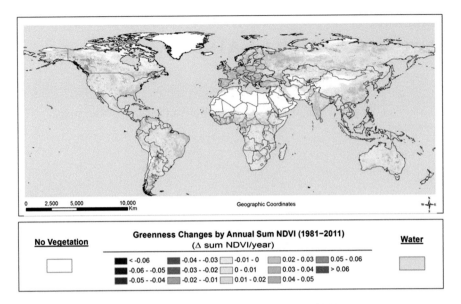

Fig. C.1 Global changes in greenness by annual sum NDVI, 1981–2011

Chaco, Pampas, and Patagonia; Southeast Asia; the grain belt from the Ukraine eastwards through Russia to Kazakhstan; the Russian far east and northeast China; and swaths of high-latitude forest.

- *All kinds of land use are afflicted.* Cropland comprises 13 % of the global land area but makes up 15 % of the total degrading land; rangeland makes up 29 % of the land but 42 % of degraded land; forest is also overrepresented, occupying 23 % of the land area but 37 % of the degrading area.
- *Comparison of rural population density with land degradation shows no simple pattern.* Taking infant mortality and the percentage of young children who are underweight as proxies for poverty, there is some correlation, but we need a more rigorous analysis.
- *Fourteen percent of the land surface has been improving over the period.*

Changing Trends

The longer time series reveals significant changes in trend over the last 30 years. Linear trend analysis is a blunt instrument, but using the Breaks for Additive Seasonal and Trend (BFAST) algorithm to analyze changes of trend de Jong et al. 2011a found that most parts of the world have experienced periodic changes of trend, even reversals. This is important for interpretation of longer-term trends where diverging trends may balance out, for instance, in China where a significant change in direction was identified around 1996. Figure C.2 illustrates persistent and

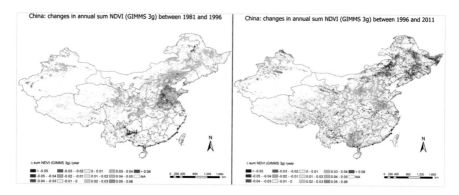

Fig. C.2 China with provincial boundaries: changes in annual sum NDVI 1981–1996 (**a**) and 1996–2011 (**b**)

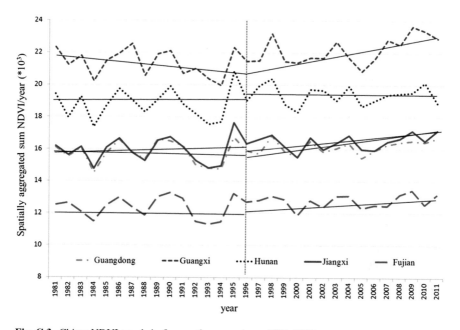

Fig. C.3 China: NDVI trends in five southern provinces 1981–2011

expanding degradation in Tibet and the southwestern provinces, a dramatic increase in degradation across the northeast, and a loss of impetus in many intensively farmed areas in spite of the increasing application of synthetic fertilizer from 7 million tons in 1977 to more than 58 million tons in 2012. Over the period 1981–1996, 1.8 % of the country suffered degradation, but 17.5 % was improving (80.7 % showed no significant change or was barren): between 1996 and 2011, 12.6 % of the land was degrading and only 10 % showed improvement (77.4 % no change or barren).

In more detail, trend analysis for the southern provinces of Guangdong, Guangxi, Hunan, Jiangxi, and Fujian, which exhibit a general improvement over the last 30 years, shows cyclically declining NDVI at the beginning of the time series but a reversal of the trends after about 1995 (Fig. C.3) which may be attributed to the take-up of the Grain-for-Green initiative (Cao et al. 2009; Bai and Dent 2014).

Chinese researchers have also made use of the GIMMS NDVI database and the GLADA methodology for several countrywide studies. In the process, some new indices have been developed, such as the sensitivity index—the degree of reaction of NDVI to rainfall change in specific rainfall regions was developed (see Appendix D).

Making Allowance for Terrain, Soil, and Land Use

We might expect resilience against land degradation to depend on terrain, the kind of soils, land use, and management. Bai and Dent (2014) used the SRTM digital elevation model and ChinaSOTER at scale 1:1 million to assess the effects of soils and terrain on land degradation and improvement. Figure C.4 depicts the relative departure of each pixel from the trend of its SOTER unit, separating the impact of soil and terrain from other factors and giving a picture of land degradation and

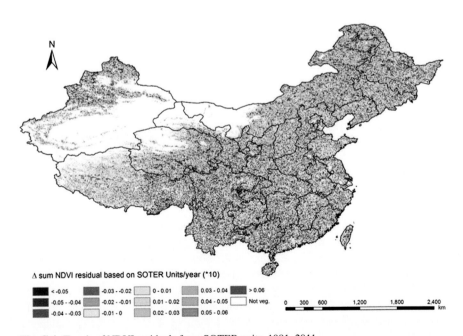

Fig. C.4 Trends of NDVI residuals from SOTER units, 1981–2011

improvement at the landscape scale—where most land-use and management decisions are taken.

This analysis examines land degradation and improvement at the landscape scale, where most land-use decisions are actually taken, and shows which parts of every landscape are doing better and which are doing worse than the landscape as a whole—so it is a pointer to specific places where remedial action may be needed. Using this technique with finer-resolution data such as MODIS NDVI would be useful for national-level reporting and for assessment of policy and project impacts on the ground.

Analysis of land degradation in terms of land use and management is more problematic because no two consecutive land-use surveys have used the same classification. Our only recourse is to historical analysis of land use change making use of Landsat imagery, e.g. Bai et al. (2010).

Appendix D
China's Experiences on the Usefulness of GLADA

Courtesy of Zhang Kebin, Beijing Forestry University, Key Authority implementing LADA in China and the PRC-GEF Land Degradation Partnership.

Introduction

Based on the practice and experience of the LADA program, the choice of the GLADA methodology as the standard method for global assessment of drylands is a good decision. The bedrock of the GLADA approach is the use of NDVI as the basic indicator, enabling consistency and comparability. Assessment methods for land degradation at the national level (LADA-national) have adopted a layered approach whereby a comprehensive analysis incorporating social information can be conducted on various environments so as to broaden the perspective of people when observing land degradation.

Rightly embracing the fundamentals of GLADA and the LADA-national approach, GLADIS leads to the improvement and enhancement of GLADA. From this point of view, GLADIS is applicable to and ready for the assessment on global land degradation and, also, at the national level for countries with a large land area, including China. Of course, at the national level, the remote sensing data should be at an appropriate resolution.

Certainly, the methodology needs constant improvement so as to make the assessment more realistic. For example, according to the GLADA assessment, since the early 1980s, subsequent years have witnessed land degradation both in arid areas and, also, increasing appearance of hot spots of degradation in more humid areas. This is the case in China, but we cannot tell whether it is a coincidence or the result of methodological problems.

NDVI is affected by various factors such as climate (rainfall), soil, kind of tillage, and so on. Fluctuations in climate (especially those of rainfall) often lead to fluctuations in NDVI. Especially in arid areas, a high sensitivity of vegetation to rainfall always results in fluctuations of NDVI according to those of rainfall. In this case, NDVI changes do not mean that the quality or long-lasting production poten-

© The Author(s) 2015
G.T. Yengoh et al., *Use of the Normalized Difference Vegetation Index (NDVI) to Assess Land Degradation at Multiple Scales*, SpringerBriefs in Environmental Science, DOI 10.1007/978-3-319-24112-8

tial of land has changed accordingly; that is to say, the land quality may exhibit no degradation or improvement. So, if we rely solely on the change of NDVI as an indicator of land degradation, such judgment possibly may draw out wrong assessment conclusions. So to eliminate the interference of rainfall fluctuations, GLADA employs the concept of RUE.

When testing the GLADA assessment against China's actual situation and an analysis of the impact of rainfall on NDVI, it is found that the actual situation is complicated; rainfall has a great impact on NDVI, but this impact is not homogeneous—it reduces gradually as rainfall increases. In fact, with rainfall reaching a certain level, the impact of increasing rainfall on NDVI gradually weakens until, in the humid area with abundant rainfall, the impact of fluctuation of rainfall on the change in NDVI becomes very small, even negligible. If RUE is still used as an indicator of land degradation in these humid areas, then we are not certain that it is appropriate. [Editor's comment: There appears to be some misunderstanding here. GLADA uses RUE to mask drought effects only in those areas where there is a direct correlation between NDVI and rainfall—essentially this applies to drylands. In humid areas (and irrigated areas), NDVI is used unmodified as a proxy for biological productivity.]

For this consideration, combining with the actual situation in China, and through an analysis of the relation between NDVI and rainfall, we have tried to determine the quantitative impact of rainfall on NDVI under different rainfall conditions (i.e., different regions), so as to use different corrective factors for different areas according to actual rainfall when eliminating the impact of rainfall on NDVI, so that it is more scientific to use NDVI as an assessment indicator for land degradation and also to supplement the GLADA method.

Data

GIMMS-8 km-NDVI data: The Chinese team used the GIMMS 2g dataset of NOAA-AVHRR NDVI 15-day synthesized data from July 1981 to September 2006 (a total of 26 years). The average value of the two fortnightly data points was taken as a monthly value, and the annual accumulated NDVI value is the sum of 12 monthly values. Data resolution is 8 km.

Rainfall data: The rainfall data come from the China National Meteorological Information Center which stores and shares historical and real-time meteorological data; all data passing a quality test are recompiled. The rainfall data used in this study comprised 26 years measured data from 707 meteorological stations in mainland China from 1981 to 2006, matching the time period of NDVI data.

The Relation Between NDVI and Rainfall

Figure D.1 shows the relationship between annual sum NDVI and annual rainfall observed from 707 meteorological stations in China from 1981 to 2006:

- There is a close relation between NDVI and rainfall.
- The relation between NDVI and rainfall is not homogeneous but nonetheless direct.
- In areas with low rainfall, the rainfall has the strongest impact on NDVI and the data are tightly aligned around the best-fitted curve. With increasing rainfall, the impact of rainfall on NDVI falls away and the correlation also gradually weakens.

Definition and calculation of the Sensitivity Index: In order to express this close relation between NDVI and rainfall, we have derived a sensitivity index of NDVI to rainfall (N_s) (in short, Sensitivity Index). The Chinese team also defined the sensitivity index as the rate of NDVI change with rainfall change by sensitivity index of NDVI to rainfall in the following mathematical expression:

$$N_s = \frac{0.0424 - 0.01313 \times \log_{10} NR_a - 0.01176 \times \log_{10}^2 NR_a}{NR_a \times \ln 10}$$

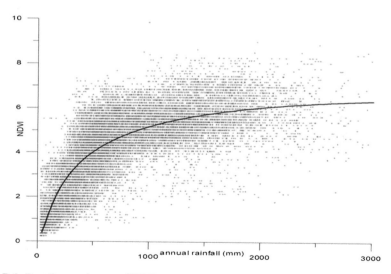

Fig. D.1 Best-fit curve between NDVI accumulated value and annual rainfall

The Sensitivity Index quantitatively reveals the rate of change of NDVI with the change of rainfall under any rainfall conditions. More importantly, the Sensitive Index may be used as the parameter for eliminating the impact of rainfall on NDVI in the assessment of land degradation.

Future Studies

More studies should be carried out on the relationship of NDVI and rainfall.

Appendix E
Main Features of Image Products from the Different Sensors

Products (sensors)	Features	Vegetation mapping applications
Landsat TM	Medium to coarse spatial resolution with multispectral data (120 m for thermal infrared band and 30 m for multispectral bands) from Landsat 4 and 5 (1982 to present). Each scene covers an area of 185×185 km. Temporal resolution is 16 days	Regional-scale mapping, usually capable of mapping vegetation at community level
Landsat ETM+ (Landsat 7)	Medium to coarse spatial resolution with multispectral data (15 m for panchromatic band, 60 m for thermal infrared, and 30 m for multispectral bands) (1999 to present). Each scene covers an area of 185 km×185 km. Temporal resolution is 16 days	Regional-scale mapping, usually capable of mapping vegetation at community level, some dominant species can possibly be discriminated
SPOT	A full range of medium spatial resolutions from 20 m down to 2.5 m and SPOT VGT with coarse spatial resolution of 1 km. Each scene covers 60×60 km for HRV/HRVIR/HRG and 1000×1000 km (or 2000×2000 km) for VGT. SPOT 1, 2, 3, 4, and 5 were launched in 1986, 1990, 1993, 1998, and 2002, respectively. SPOT 1 and 3 are not providing data now	Regional scale, usually capable of mapping vegetation at community level or species level, global-/national-/regional-scale (from VGT) mapping land-cover types (i.e., urban area, classes of vegetation, water area, etc.)
MODIS	Coarse spatial resolution (250–1000 m) and multispectral data from the Terra Satellite (2000 to present) and Aqua Satellite (2002 to present). Revisit interval is 1–2 days. The swath is 2330 km (cross track) by 10 km (along track at nadir)	Mapping at global, continental, or national scale. Suitable for vegetation mapping at a coarse scale and land-cover types (urban area, classes of vegetation, water area, etc.)

(continued)

© The Author(s) 2015
G.T. Yengoh et al., *Use of the Normalized Difference Vegetation Index (NDVI) to Assess Land Degradation at Multiple Scales*, SpringerBriefs in Environmental Science, DOI 10.1007/978-3-319-24112-8

(continued)

Products (sensors)	Features	Vegetation mapping applications
AVHRR	1-km GSD with multispectral data from the NOAA satellite series (1980 to present). The approximate scene size is 2400 × 6400 km	Global-, continental-, or national-scale mapping. Suitable for mapping land-cover types (urban area, classes of vegetation, water area, etc.)
IKONOS	High-resolution imagery at 1-m (panchromatic) and 4-m (multispectral bands, including red, green, blue, and near-infrared) resolution. The revisit rate is 3–5 days (off-nadir). The single scene is 11 × 11 km	Local- to regional-scale vegetation mapping at species or community level. Can be used to validate other classification results
QuickBird	High-resolution (2.4–0.6 m) panchromatic and multispectral imagery from a constellation of spacecraft. Single scene area is 16.5 × 16.5 km. Revisit frequency is around 1–3.5 days depending on latitude	Local- to regional-scale vegetation mapping at species or community level. Can be used to validate vegetation cover intepreted from other images
ASTER	Medium spatial resolution (15–90 m) with 14 spectral bands from the Terra Satellite (2000 to present). Visible to near-infrared bands have a spatial resolution of 15 m, 30 m for shortwave infrared bands, and 90 m for thermal infrared bands	Regional- to national-scale vegetation mapping at species or community level
AVIRIS	Airborne sensor collecting images with 224 spectral bands from visible, near infrared to shortwave infrared. Depending on the satellite platforms and latitude of data collected, the spatial resolution ranges from meters to dozens of meters and the swath ranges from several kilometers to dozens of kilometers	At local to regional scale usually capable of mapping vegetation at community level or even species level. Images are created as one-time operations on an 'as needs' basis, data are not readily available.
Hyperion	Hyperspectral image with 220 bands ranging from visible to shortwave infrared. The spatial resolution is 30 m. Data available since 2003	At regional scale, capable of mapping vegetation at community level or species level

Source: Modified from (Xie et al. 2008)

Appendix F
UNCCD Core Indicators
for National Reporting: ICCD/COP(11)/CST/2

Proposed refinements to the provisionally adopted set of impact indicators

Indicator	Metrics/proxies	Description	Potential data source/ reference methodology
Strategic Objective 1: to improve the living conditions of affected populations			
Trends in population living below the relative poverty line and/or income inequality in affected areas	Poverty severity (or squared poverty gap)	Takes account of both the distance separating the poor from the poverty line and the inequality among the poor	World Bank[a,b]
	or		
	Income inequality	Alternative to the poverty severity metric for those countries where poverty is no longer an issue; Strategic Objective 1 has, in this sense, already been achieved	OECD[c]
Trends in access to safe drinking water in affected areas	Proportion of population using an improved drinking water source	An improved drinking water source is defined as one that is protected from outside contamination through household connection, public standpipe, borehole, protected dug well, protected spring, rainwater, etc.	WHO/UNICEF Joint Monitoring Program for Water Supply and Sanitation methodology[d]

(continued)

© The Author(s) 2015
G.T. Yengoh et al., *Use of the Normalized Difference Vegetation Index (NDVI) to Assess Land Degradation at Multiple Scales*, SpringerBriefs in Environmental Science, DOI 10.1007/978-3-319-24112-8

(continued)

Proposed refinements to the provisionally adopted set of impact indicators

Indicator	Metrics/proxies	Description	Potential data source/ reference methodology
Strategic Objective 2: to improve the condition of ecosystems			
Trends in land-cover structure	Vegetative land-cover structure	Intended as the distribution of land-cover types of greatest concern for land degradation (excluding artificial surfaces) by characterizing the spatial structure of vegetative land cover; it should include and specify natural habitat classes	Sourced from products like GlobCover*e,f* or finer-resolution products under development (Gong et al. 2013), and following established land-cover classifications (e.g., FAO/UNEP LCCS[g])
Trends in land productivity or functioning of the land	Land-productivity dynamics	Based on long-term fluctuations and current efficiency levels of phenology and productivity factors affecting standing biomass conditions	New World Atlas of Desertification methodology;[h] update foreseen every 5 years
Strategic Objective 3: to generate global benefits through effective implementation of the UNCCD			
Trends in carbon stocks above- and belowground	Soil organic carbon stock	Intended as the status of topsoil and subsoil organic carbon	Sourced from, e.g., the GTOS portal[i]
	To be replaced by		
	Total terrestrial system carbon stock	Including above- and below-ground carbon	To be streamlined with the GEF-financed UNEP Carbon Benefits Project[j,k]
	Once operational		
Trends in abundance and distribution of selected species (*potentially to be replaced by an indicator measuring trends in ecosystem functional diversity once system understanding and data allows*)	Global Wild Bird Index	Measures average population trends of a suite of representative wild birds, as an indicator of the general health of the wider environment	Following the indicator guidance provided for and to be streamlined with the CBD process[l,m]

Abbreviations: *CBD* convention on biological diversity, *FAO* food and agriculture organization of the United Nations, *GEF* global environment facility, *GTOS* global terrestrial observing system, *LCCS* land cover classification system, *OECD* organisation for economic co-operation and development, *UNEP* United Nations environment programme, *UNICEF* United Nations children's fund, *WHO* world health organization

[a]http://web.worldbank.org/WBSITE/EXTERNAL/TOPICS/EXTPOVERTY/EXTPA/0,contentM DK:20242881~isCURL:Y~menuPK:492130~pagePK:148956~piPK:216618~theSit ePK:430367,00.html
[b]http://siteresources.worldbank.org/INTPA/Resources/tn_measuring_poverty_over_time.pdf
[c]http://www.oecd.org/els/soc/43540354.pdf
[d]http://www.wssinfo.org/
[e]http://due.esrin.esa.int/globcover/
[f]http://www.gofcgold.wur.nl/sites/gofcgold_refdataportal.php
[g]http://www.fao.org/docrep/003/X0596E/X0596e00.htm
[h]http://wad.jrc.ec.europa.eu/
[i]http://www.fao.org/gtos/tcoDAT.html
[j]http://carbonbenefitsproject-compa.colostate.edu/
[k]http://www.unep.org/climatechange/carbon-benefits/Home/tabid/3502/Default.aspx
[l]http://www.unep-wcmc.org/wild-bird-index_568.html
[m]http://www.bipindicators.net/WBI

Appendix G
Current Cost[a] of Selected Satellite Imagery

© The Author(s) 2015
G.T. Yengoh et al., *Use of the Normalized Difference Vegetation Index (NDVI) to Assess Land Degradation at Multiple Scales*, SpringerBriefs in Environmental Science, DOI 10.1007/978-3-319-24112-8

Type	Years	Bands	MS (m)	Pan (m)	Scene size (km)	Cost (km^2)
CORONA http://www.nro.gov/history/csnr/corona/factsheet.html	1960–1972	1	–	08-Feb	17×232	$30 per scene
Landsat 4–5 MSS http://landsat.usgs.gov/about_landsat5.php	1982–1999	4	80	–	170×185	Free
Landsat 4–5 TM http://landsat.usgs.gov/about_landsat5.php	1982–2012	7	30	–	170×185	Free
Landsat 7 ETM+ http://landsat.usgs.gov/about_landsat7.php	1999–present	8	30	15	170×185	Free
Landsat 8 http://landsat.usgs.gov/landsat8.php	2013–present	8	30	15	170×185	Free
GIMMS and GIMMS3g http://ecocast.arc.nasa.gov/data/pub/gimms/	1981–present	1	8000		Continuous global	Free
MODIS http://modis.gsfc.nasa.gov/data/	1999–present	36	250, 500, 1000		Continuous global	Free
SPOT 1–3 http://www.geo-airbusds.com/en/143-spot-satellite-imagery	1986–1997	3	20	10	60×60	$1200 per scene
SPOT 4 http://www.geo-airbusds.com/en/143-spot-satellite-imagery	1998–2013	4	20	10	60×60	$1200 per scene

Satellite	Period					Cost
SPOT 5 http://www.geo-airbusds.com/en/143-spot-satellite-imagery	2002–present	4	10	2.5/5	60 × 60	$2700 per scene
SPOT 6–7 http://www.geo-airbusds.com/en/143-spot-satellite-imagery	2012–present	4	6	1.5	60	$5.15
IKONOS https://www.digitalglobe.com/	1999–present	4	4	1	11.3	$10
QuickBird https://www.digitalglobe.com/	2001–2014	4	2.4	0.6	16.8	$16
Pléiades 1A-1B http://www.geo-airbusds.com/pleiades/	2011–present	4	2	0.5	20	$13
WorldView-1 https://www.digitalglobe.com	2007–present	1	–	0.46	17.6 × 14	$13
WorldView-2 https://www.digitalglobe.com	2009–present	8	1.84	0.46	16.4	$29
WorldView-3 https://www.digitalglobe.com	2014–present	28	1.24	0.31	13.1	–

MS multispectral resolution, *Pan* panchromatic resolution

Scene size the total coverage of the scene or the maximum swath width (if just a single number)

[a]This information was put together in September 2014. Prices are only estimates based on online sources, and may differ depending on sales outlets. Academic discounts can range from 20 to 30 %. http://rmseifried.com/2014/09/19/satellite-imagery-types-resolution-and-pricing/

Appendix H
Software for Processing Satellite Images to Develop the NDVI

Common Commercial Software for Processing Satellite Images to Develop the NDVI

ArcGIS. ArcGIS is the most popular geospatial product in the industry. According to its creators, the Environmental Systems Research Institute (ESRI), the ArcGIS platform is a complete GIS suite that permits users to conduct spatial analysis, manage data effectively, explore data content, automate advanced workflows, create maps, and undertake advanced operations with, and analysis of, imagery. It comprises a range of software tools and extensions (most of which are available to users depending on the license level of their platforms) with sophisticated desktop GIS functionality such as spatial analysis, 3D, geostatistics, server-based data management, and geo-processing that leverage Oracle or SQLServer type databases. While a range of operations on imagery can be achieved through the use of tools in extensions such as Spatial Analyst, the Image Analyst has functionality for performing basic tasks such as image subtraction, vegetation indices, and others. More on the features of ArcGIS for Desktop, its extensions and capabilities can be found here: http://www.esri.com/software/arcgis/arcgis-for-desktop.

© The Author(s) 2015
G.T. Yengoh et al., *Use of the Normalized Difference Vegetation Index (NDVI) to Assess Land Degradation at Multiple Scales*, SpringerBriefs in Environmental Science, DOI 10.1007/978-3-319-24112-8

Erdas Imagine. Erdas Imagine is a powerful but simple-to-use geospatial imagery processing application. It is renowned for its ability to enhance satellite imagery to make it more meaningful and facilitate complex analysis. It has built-in tools for the interchange of numerous file formats and has an extensive array of vector editing and analysis capabilities and the ability to design custom geospatial-analysis tools, handle multispectral and hyperspectral data, as well as advanced classification techniques. More on Erdas Imagine can be found here: http://www.hexagongeospatial. com/products/remote-sensing/erdas-imagine.

ER Mapper. ER Mapper is a geographic image processing software product, which runs on a range of PCs running Windows. This software can be used to display, integrate, and enhance raster data, display and edit vector data, and link with data from Geographic and Land Information Systems, Database Management Systems, or indeed most other sources. The outstanding feature and benefit of ER Mapper is its ease of use and processing on the fly feature. This means that different image-processing algorithms can be launched simultaneously, and the results instantly displayed on the user's screen. This reduces the time it takes for other applications to do the same tasks, because, generally, a new file would have to be created before the results are displayed. All processing information can be saved in project files called Algorithms where all the processing steps are on the image or images. More on ER Mapper can be found here: http://www.hexagongeospatial.com/products/remote-sensing/erdas-er-mapper.

Geospatial software for monitoring and modeling the Earth system

IDRISI. IDRISI GIS and image processing application is developed by Clark Labs at Clark University. It has been seen by many in the remote sensing community as affordable and robust and is widely used for both professional operations and education. IDRISI is currently in the Selva version. This is an integrated GIS and Image Processing software package with over 300 modules for the processing, analysis, and display of digital spatial information. It may be considered the most extensive set of GIS and Image Processing tools in the industry in a single, affordable package. It includes tools such as neural network classifiers and change detection modules that are not available out-of-the-box in any other software. TerrSet is a recent integrated geospatial software system for monitoring and modeling the earth system for sustainable development. It incorporates the IDRISI GIS Analysis and image processing tools along with a collection of vertical applications to perform complex analysis of issues of sustainability in the earth system. More on IDRISI and its associated sets of integrated tools can be found here: http://www.clarklabs. org/.

ENVI. ENVI products create the primary geospatial software foundation to convert, preprocess, process, and analyze all types of imagery and data such as multispectral, hyperspectral, LiDAR, and SAR. They are known for their design which permits easy acquaintance with and use by everyone regardless of prior experience with imagery—from GIS professionals to image analysts and image scientists. ENVI products interoperate well with many geospatial applications. For example, all ENVI products integrate with ArcGIS, and can be easily customized to meet unique needs of users across disciplines. More about ENVI and its family of products can be found here: http://www.exelisvis.com/ProductsServices/ENVIProducts/ENVI. aspx.

PCI Geomatica. PCI Geomatica is an integrated software featuring the majority of tools needed by professionals for remote sensing, digital photogrammetry, image analysis, map production, and mosaicking. A strong point is the ease of loading satellite and aerial imagery and its capability of faster data processing with advanced modules to handle complex processes. Geomatica is also known for its comprehensive routines for automation of data processing tasks. Geomatica has been and continues to be used by many educational institutions and scientific programs throughout the world for a range of remote sensing projects. More about PCI Geomatics can be found here: http://www.pcigeomatics.com.

Common Open-Source Software

Courtesy of GISGeography.com (with some modifications)[1]

SAGA GIS: *System for Automated Geoscientific Analyses*. SAGA GIS is on the top of the list of free remote sensing softare, and for good reason. SAGA GIS is ideal for most GIS and remote sensing needs, having a rich library of modules that offer possibilities for quick, reliable raster analyses and manipulation. What gives SAGA GIS a kick is its quick and reliable raster processing. The official user guide is available here: http://www.saga-gis.org/en/

Opticks. The neat part about this software is the *long list of extensions* you can add. There are plug-ins for raster math, radar processing, and hyper-/multispectral data processing. It is important to always make sure to check the compatibility before downloading an extension. You might have to scale back your Opticks version in order for the extension to work properly. For all information concerning Optiks, as well as the download of the software, go here: http://opticks.org/confluence/display/opticks/Welcome+To+Opticks.

GRASS: *Geographic Resources Analysis Support System*. GRASS may be one of the most popular software package on this list. GRASS is *full of functionality*: image classification, PCA, edge detection, radiometric corrections, 3D, geostatistics analysis, and filtering options. Another key feature of GRASS is the LiDAR processing and analysis. You can filter LiDAR points, create contours, and

---

[1] We greatly acknowledge the contribution of GISGeography.com in permitting the use of its summary of current freely available remote sensing software below (with some modifications). The original resource provided by GISGeography can be found here: http://gisgeography.com/open-source-remote-sensing-software-packages/.

generate DEMs. The official user guide is available here: https://grass.osgeo.org/grass70/manuals/.

ORFEO: Optical and Radar Federated Earth Observation. The ORFEO toolbox was a cooperative project developed by France and Italy. It is a library of remote sensing image processing *specifically aimed at high spatial resolution.* ORFEO provides a wide range of remote sensing functions: radiometry, PCA, change detection, pan sharpening, image segmentation, classification, and filtering. An interesting aspect of this software is the capability to do object-based image analysis. This is a rare feature in software nowadays. The official user guide is available here: https://www.orfeo-toolbox.org/SoftwareGuide/SoftwareGuidech33.html.

OSSIM: Open-Source Software Image Map. OSSIM is a high-performance open-source remote sensing software application for high spatial resolution imagery. It has been actively developed for almost two decades, funded through US departments such as in intelligence and defense. Key features are compatibility with more than 100 raster and vector formats and over 4000 different projections and datums. It supports numerous sensors, but some may require additional plug-ins.

InterImage. InterImage is a bit different from the other open-source remote sensing software on this list. It specializes in automatic image interpretation. The core theme of automatic image interpretation is object-based classification (OBIA). This involves segmentation, exploring attributes, and supervised classification. Although developed in Brazil, documentation is available in English. More can be found here: http://www.lvc.ele.pucrio.br/projects/interimage/

ILWIS: *Integrated Land and Water Information System.* ILWIS has been around for more than 25 years. It has had over 27,000 downloads since its first release. It wasn't until recently that it has become available for public use. ILWIS was originally built for researchers and students. For this reason, effort was concentrated on developing a user-friendly environment. The other main focus was compatibility with raster and vector formats. This has been done by full integration with the GDAL library. The practical uses of ILWIS make it a prime choice for remote sensing activities. More on ILWIS can be found here: http://www.ilwis.org/

gvSIG. gvSIG is known for its wide variety of rich features and powerful capabilities: supervised classification, defining ROIs, band algebra, and decision trees. gvSIG stands for Generalitat Valenciana Geographic Information System. Generalitat Valenciana is the Spanish regional authority the system was designed for. More on gvSIG can be found here: http://www.gvsig.com/en

QUANTUM GIS

Quantum GIS (QGIS). QGIS is one of the *most powerful open-source GIS software packages* available for free. This software package allows users to visualize, analyze, interpret, and understand spatial data. Plug-ins are the key to its operations. Raster manipulation includes neighborhood analysis, map algebra, surface interpolation, hydrologic modeling, and terrain analysis like slope and aspect. There are plug-ins for, among other operations, semiautomated classification, BEAM and NEST framework, multitemporal raster analysis, viewshed analysis. The official site of QGIS is found here: http://www.qgis.org/en/site/.

References

Achard F, DeFries R, Eva H, Hansen M, Mayaux P, Stibig H (2007) Pan-tropical monitoring of deforestation. Environ Res Lett 2(4):045022

Adger WN (2000) Social and ecological resilience: are they related? Prog Hum Geogr 24(3):347–364

Ai L, Fang N, Zhang B, Shi Z (2013) Broad area mapping of monthly soil erosion risk using fuzzy decision tree approach: integration of multi-source data within GIS. Int J Geogr Inf Sci 27(6):1251–1267

Albalawi EK, Kumar L (2013) Using remote sensing technology to detect, model and map desertification: a review. J Food Agric Environ 11:791–797

Albergel J (1988) Genèse et prédétermination des crues au Burkina Faso: du m2 au km2: étude des paramètres hydrologiques et de leur évolution. Ph.D. Thesis. Editions de l'Orstom, 1988, 330p, Université Paris 6

Anyamba A, Tucker C (2005) Analysis of Sahelian vegetation dynamics using NOAA-AVHRR NDVI data from 1981–2003. J Arid Environ 63(3):596–614

Anyamba A, Tucker CJ (2012) Historical perspective of AVHRR NDVI and vegetation drought monitoring. Remote Sens Drought: Innovative Monit Approaches 23

Bai Z, Dent D (2014) Mapping soil degradation by NDVI, vol 2, Encyclopedia of soil science. Taylor & Francis, New York

Bai ZG, Dent DL, Olsson L, Schaepman ME (2008) Proxy global assessment of land degradation. Soil Use Manag 24(3):223–234

Bai Z, Dent D, Wu Y, de Jong R (2013) Land degradation and ecosystem services. In: Lal R, Lorenz K, Hüttl RF, Schneider BU, von Braun J (eds) Ecosystem services and carbon sequestration in the biosphere. Springer, Dordrecht, pp 357–381

Bajocco S, De Angelis A, Perini L, Ferrara A, Salvati L (2012) The impact of land use/land cover changes on land degradation dynamics: a Mediterranean case study. Environ Manage 49(5):980–989

Bandyopadhyay N, Saha AK (2014) Analysing meteorological and vegetative drought in Gujarat. In: Singh M, Singh RB, Hassan I (eds) Climate change and biodiversity: proceedings of IGU Rohtak conference, vol 1. Springer, Japan, pp 61–71

Bartholomé E, Belward A (2005) GLC2000: a new approach to global land cover mapping from Earth observation data. Int J Remote Sens 26(9):1959–1977

Beck P, Karlsen S, Skidmore A, Nielsen L, Høgda K (2005) The onset of the growing season in northwestern Europe, mapped using MODIS NDVI and calibrated using phenological ground

© The Author(s) 2015

G.T. Yengoh et al., *Use of the Normalized Difference Vegetation Index (NDVI) to Assess Land Degradation at Multiple Scales*, SpringerBriefs in Environmental Science, DOI 10.1007/978-3-319-24112-8

observations. In: 31st International Symposium on remote Sensing on Environment–Global Monitoring for Sustainability and Security, pp 20–24

Beck HE, McVicar TR, van Dijk AI, Schellekens J, de Jeu RA, Bruijnzeel LA (2011) Global evaluation of four AVHRR–NDVI data sets: intercomparison and assessment against Landsat imagery. Remote Sens Environ 115(10):2547–2563

Bennett R, Welham K, Hill RA, Ford AL (2012) The application of vegetation indices for the prospection of archaeological features in grass-dominated environments. Archaeol Prospect 19(3):209–218

Bernoux M, Chevallier T (2014) Carbon in dryland soils—Multiple essential functions, 10th edn, Les dossiers thématiques du CSFD. CSFD/Agropolis International, Montpellier, p 40 pp

Bierman R, Stocking M, Bouwman H, Cowie A, Diaz S, Granit J, Patwardhan A, Sims R, Duron G, Gorsevski V, Hammond T, Wellington-Moore C (2014) Delivering global environmental benefits for sustainable development. Report of the Scientific and Technical Advisory Panel (STAP) to the 5th GEF Assembly, México 2014. Global Environment Facility, Washington, DC

Birth GS, McVey GR (1968) Measuring the color of growing turf with a reflectance spectrophotometer. Agron J 60(6):640–643

Bradley BA, Mustard JF (2008) Comparison of phenology trends by land cover class: a case study in the Great Basin, USA. Glob Chang Biol 14(2):334–346

Brandt M, Mbow C, Diouf AA, Verger A, Samimi C, Fensholt R (2014) Ground and satellite based evidence of the biophysical mechanisms behind the greening Sahel. Global change biology 21(4):1610–1620

Cao M, Prince SD, Li K, Tao B, Small J, Shao X (2003) Response of terrestrial carbon uptake to climate interannual variability in China. Glob Chang Biol 9(4):536–546

Cao S, Chen L, Liu Z (2009) An investigation of Chinese attitudes toward the environment: case study using the Grain for Green Project. AMBIO 38(1):55–64

CEISIN (2004) Global Rural–urban mapping project: urban/rural extents. Center for International Earth Science Information Network, Columbia University, Palisades

CEISIN (2007) Socioeconomic data applications center. Center for International Earth Science Information Network, Columbia University, Palisades

Cepicky J, Becchi L (2007) Geospatial processing via internet on remote servers-PyWPS. OSGeo J 1(5):39–42

Chen D, Huang J, Jackson TJ (2005) Vegetation water content estimation for corn and soybeans using spectral indices derived from MODIS near- and short-wave infrared bands. Remote Sens Environ 98(2–3):225–236

Chen T, Niu RQ, Li PX, Zhang LP, Du B (2011) Regional soil erosion risk mapping using RUSLE, GIS, and remote sensing: a case study in Miyun Watershed, North China. Environ Earth Sci 63(3):533–541

Cherlet M, Ivits E, Sommer S, Tóth G, Jones A, Montanarella L, Belward A (2013) Land-productivity dynamics in Europe. Towards valuation of land degradation in the European Union. European Commission, Joint Research Centre, Ispra. http://wad.jrc.ec.europa.eu/data/EPreports/LPDinEU_final_no-numbers.pdf. Accessed 17 Dec 2014

Chuvieco E, Cocero D, Riano D, Martin P, Martınez-Vega J, de la Riva J, Pérez F (2004) Combining NDVI and surface temperature for the estimation of live fuel moisture content in forest fire danger rating. Remote Sens Environ 92(3):322–331

Conijn J, Bai Z, Bindraban P, Rutgers B (2013) Global changes of net primary productivity, affected by climate and abrupt land use changes since 1981. Towards mapping global soil degradation. Report 2013/01, ISRIC–World Soil Information, Wageningen. ISRIC Report 1

Cook BI, Pau S (2013) A global assessment of long-term greening and browning trends in pasture lands using the GIMMS LAI3g dataset. Remote Sens 5(5):2492–2512

Cowie A, Penman T, Gorissen L, Winslow M, Lehmann J, Tyrrell T, Twomlow S, Wilkes A, Lal R, Jones J (2011) Towards sustainable land management in the drylands: scientific connections in monitoring and assessing dryland degradation, climate change and biodiversity. Land Degrad Dev 22(2):248–260

Cui X, Gibbes C, Southworth J, Waylen P (2013) Using remote sensing to quantify vegetation change and ecological resilience in a semi-arid system. Land 2(2):108–130

Dardel C, Kergoat L, Hiernaux P, Mougin E, Grippa M, Auda Y, Tucker CJ (2013) 30 years of remote sensing imagery in Sahel confronted to field observations (Gourma, Mali). In: EGU General Assembly Conference Abstracts, p 12872

Dardel C, Kergoat L, Hiernaux P, Grippa M, Mougin E, Ciais P, Nguyen C-C (2014) Rain-use-efficiency: what it tells us about the conflicting Sahel greening and Sahelian Paradox. Remote Sens 6(4):3446–3474

De Angelis A, Bajocco S, Ricotta C (2012) Modelling the phenological niche of large fires with remotely sensed NDVI profiles. Ecol Model 228:106–111

De Beurs K, Henebry G (2005) A statistical framework for the analysis of long image time series. Int J Remote Sens 26(8):1551–1573

de Jong R (2010) Trends in soil degradation publications. IUSS Bulletin 116:21–25

de Jong R, de Bruin S, de Wit A, Schaepman ME, Dent DL (2011a) Analysis of monotonic greening and browning trends from global NDVI time-series. Remote Sens Environ 115(2):692–702. doi:http://dx.doi.org/10.1016/j.rse.2010.10.011

de Jong R, de Bruin S, Schaepman M, Dent D (2011b) Quantitative mapping of global land degradation using Earth observations. Int J Remote Sens 32(21):6823–6853

de Jong R, Verbesselt J, Schaepman ME, De Bruin S (2011c) Detection of breakpoints in global NDVI time series. In: 34th International Symposium on Remote Sensing of Environment (ISRSE), pp 10–15

Dee D, Uppala S, Simmons A, Berrisford P, Poli P, Kobayashi S, Andrae U, Balmaseda M, Balsamo G, Bauer P (2011) The ERA-interim reanalysis: configuration and performance of the data assimilation system. Q J Roy Meteorol Soc 137(656):553–597

DeFries R, Townshend J (1994) NDVI-derived land cover classifications at a global scale. Int J Remote Sens 15(17):3567–3586

Delbart N, Kergoat L, Le Toan T, Lhermitte J, Picard G (2005) Determination of phenological dates in boreal regions using normalized difference water index. Remote Sens Environ 97(1):26–38

Dent D (2007) Environmental geophysics mapping salinity and water resources. Int J Appl Earth Obs Geoinf 9(2):130–136

Dent D, Bai Z, Schaepman M, Olsson L (2009) Letter to the editor. Soil Use Manag 25(1):93–97

Despland E, Rosenberg J, Simpson SJ (2004) Landscape structure and locust swarming: a satellite's eye view. Ecography 27(3):381–391

Di Gregorio A (2005) Land cover classification system: classification concepts and user manual: LCCS, vol 8, 8th edn, FAO environment and natural resources service series. Food and Agriculture Organization of the United Nations, Rome

Díaz-Delgado R, Lloret F, Pons X, Terradas J (2002) Satellite evidence of decreasing resilience in Mediterranean plant communities after recurrent wildfires. Ecology 83(8):2293–2303

Díaz-Delgado R, Lloret F, Pons X (2003) Influence of fire severity on plant regeneration by means of remote sensing imagery. Int J Remote Sens 24(8):1751–1763

Diouf A, Lambin E (2001) Monitoring land-cover changes in semi-arid regions: remote sensing data and field observations in the Ferlo, Senegal. J Arid Environ 48(2):129–148

Duro DC, Coops NC, Wulder MA, Han T (2007) Development of a large area biodiversity monitoring system driven by remote sensing. Prog Phys Geogr 31(3):235–260

Eldeiry A, Garcia L (2010) Comparison of ordinary kriging, regression kriging, and cokriging techniques to estimate soil salinity using LANDSAT images. J Irrig Drain Eng 136(6):355–364. doi:10.1061/(ASCE)IR.1943-4774.0000208

Erian WF (2005) Arab network of the remote sensing centers for desertification monitoring and assessment. ed. In: Remote sensing and geoinformation processing in the assessment and monitoring of land degradation and desertification. Trier, Germany, pp 452–459

Evans J, Geerken R (2004) Discrimination between climate and human-induced dryland degradation. J Arid Environ 57(4):535–554

FAO (2009) High level expert forum—how to feed the world in 2050. Office of the Director, Agricultural Development Economics Division, Economic and Social Development Department, Rome

FAO (2013) Land use systems of the world. Land cover and land use. Food and Agriculture Organization of the United Nations (FAO), Rome. http://www.fao.org/climatechange/54270/en/. Accessed 15 June 2013

Farifteh J, Farshad A, George R (2006) Assessing salt-affected soils using remote sensing, solute modelling, and geophysics. Geoderma 130(3):191–206

Fensholt R, Proud SR (2012) Evaluation of Earth Observation based global long term vegetation trends—Comparing GIMMS and MODIS global NDVI time series. Remote Sens Environ 119:131–147. doi:http://dx.doi.org/10.1016/j.rse.2011.12.015

Fensholt R, Rasmussen K (2011) Analysis of trends in the Sahelian 'rain-use efficiency' using GIMMS NDVI, RFE and GPCP rainfall data. Remote Sens Environ 115(2):438–451

Fensholt R, Langanke T, Rasmussen K, Reenberg A, Prince SD, Tucker C, Scholes RJ, Le QB, Bondeau A, Eastman R, Epstein H, Gaughan AE, Hellden U, Mbow C, Olsson L, Paruelo J, Schweitzer C, Seaquist J, Wessels K (2012) Greenness in semi-arid areas across the globe 1981–2007—an Earth Observing Satellite based analysis of trends and drivers. Remote Sens Environ 121:144–158. doi:10.1016/j.rse.2012.01.017

Fensholt R, Rasmussen K, Kaspersen P, Huber S, Horion S, Swinnen E (2013) Assessing land degradation/recovery in the African Sahel from long-term earth observation based primary productivity and precipitation relationships. Remote Sens 5(2):664–686

Field CB, Randerson JT, Malmström CM (1995) Global net primary production: combining ecology and remote sensing. Remote Sens Environ 51(1):74–88

Foley JA, Ramankutty N, Brauman KA, Cassidy ES, Gerber JS, Johnston M, Mueller ND, O'Connell C, Ray DK, West PC (2011) Solutions for a cultivated planet. Nature 478:337–342

Foth HD (1991) Fundamentals of soil science, vol Ed. 8. Wiley, New York

Friedl MA, McIver DK, Hodges JC, Zhang X, Muchoney D, Strahler AH, Woodcock CE, Gopal S, Schneider A, Cooper A (2002) Global land cover mapping from MODIS: algorithms and early results. Remote Sens Environ 83(1):287–302

Fu B, Liu Y, Lü Y, He C, Zeng Y, Wu B (2011) Assessing the soil erosion control service of ecosystems change in the Loess Plateau of China. Ecol Complex 8(4):284–293

Gallo K, Ji L, Reed B, Eidenshink J, Dwyer J (2005) Multi-platform comparisons of MODIS and AVHRR normalized difference vegetation index data. Remote Sens Environ 99(3):221–231

Gao B-C (1996) NDWI—a normalized difference water index for remote sensing of vegetation liquid water from space. Remote Sens Environ 58(3):257–266

GEF (2014) EF-6 programming directions. GEF/R.6/20/Rev.04, 31 March 2014. Global Environment Facility, Washington, DC

Gentemann CL, Wentz FJ, Mears CA, Smith DK (2004) In situ validation of Tropical Rainfall Measuring Mission microwave sea surface temperatures. J Geophys Res: Oceans (1978–2012) 109(C4):C04021

Gibbes C, Southworth J, Waylen P, Child B (2014) Climate variability as a dominant driver of post-disturbance savanna dynamics. Appl Geogr 53:389–401

Gilabert M, González-Piqueras J, Garcıa-Haro F, Meliá J (2002) A generalized soil-adjusted vegetation index. Remote Sens Environ 82(2):303–310

Goetz SJ, Prince SD, Goward SN, Thawley MM, Small J (1999) Satellite remote sensing of primary production: an improved production efficiency modeling approach. Ecol Model 122(3):239–255

Gottwald M, Bovensmann H, Lichtenberg G, Noel S, von Bargen A, Slijkhuis S, Piters A, Hoogeveen R, von Savigny C, Buchwitz M (2006) SCIAMACHY, monitoring the changing Earth's atmosphere. DLR, Germany

Goward SN, Huemmrich KF (1992) Vegetation canopy PAR absorptance and the normalized difference vegetation index: an assessment using the SAIL model. Remote Sens Environ 39(2):119–140

Goward SN, Tucker CJ, Dye DG (1985) North American vegetation patterns observed with the NOAA-7 advanced very high resolution radiometer. Vegetatio 64(1):3–14

Grainger A (2013) The threatening desert: controlling desertification. Routledge, London

Green RM, Hay SI (2002) The potential of Pathfinder AVHRR data for providing surrogate climatic variables across Africa and Europe for epidemiological applications. Remote Sens Environ 79(2):166–175

Haboudane D, Miller JR, Pattey E, Zarco-Tejada PJ, Strachan IB (2004) Hyperspectral vegetation indices and novel algorithms for predicting green LAI of crop canopies: modeling and validation in the context of precision agriculture. Remote Sens Environ 90(3):337–352

Haigh MJ (2002) Land use, land cover and soil sciences—land rehabilitation, vol IV, Encyclopedia of life support systems (EOLSS). UNESCO-EOLSS, Oxford

Hansen M, DeFries R, Townshend JR, Sohlberg R (2000) Global land cover classification at 1 km spatial resolution using a classification tree approach. Int J Remote Sens 21(6–7):1331–1364

Hassan R, Scholes R, Ash N (2005) Ecosystems and human well-being: current state and trends, vol 1. Island Press, Washington, DC

Hayward MW, Zawadzka B (2010) Increasing elephant Loxodonta Africana density is a more important driver of change in vegetation condition than rainfall. Acta Theriol 55(4):289–298

Haywood J, Randall J (2008) Trending seasonal data with multiple structural breaks. NZ visitor arrivals and the minimal effects of 9/11. Research report 08/10. Victoria, New Zealand, University of Wellington. Vol. p 26

He KS, Zhang J, Zhang Q (2009) Linking variability in species composition and MODIS NDVI based on beta diversity measurements. Acta Oecol 35(1):14–21

Henricksen B, Durkin J (1986) Growing period and drought early warning in Africa using satellite data. Int J Remote Sens 7(11):1583–1608

Herrmann SM, Sop T (2015) The map is not the territory. How satellite remote sensing and ground evidence have (Re-)Shaped the image of Sahelian desertification. In: Behnke R, Mortimore M (eds) Desertification: science, politics and public perception. Springer (Earth System Science Series), New York

Herrmann SM, Tappan GG (2013) Vegetation impoverishment despite greening: a case study from central Senegal. J Arid Environ 90:55–66

Herrmann SM, Anyamba A, Tucker CJ (2005) Recent trends in vegetation dynamics in the African Sahel and their relationship to climate. Glob Environ Chang 15(4):394–404

Herrmann SM, Sall I, Sy O (2014) People and pixels in the Sahel: a study linking coarse-resolution remote sensing observations to land users' perceptions of their changing environment in Senegal. Ecol Soc 19(3):29

Hickler T, Eklundh L, Seaquist JW, Smith B, Ardö J, Olsson L, Sykes MT, Sjöström M (2005) Precipitation controls Sahel greening trend. Geophys Res Lett 32(21)

Higginbottom TP, Symeonakis E (2014) Assessing land degradation and desertification using vegetation index data: current frameworks and future directions. Remote Sens 6(10):9552–9575

Hilker T, Lyapustin AI, Tucker CJ, Hall FG, Myneni RB, Wang Y, Bi J, de Moura YM, Sellers PJ (2014) Vegetation dynamics and rainfall sensitivity of the Amazon. Proc Natl Acad Sci 111(45):16041–16046

Hoekstra JM, Boucher TM, Ricketts TH, Roberts C (2005) Confronting a biome crisis: global disparities of habitat loss and protection. Ecol Lett 8(1):23–29

Holben BN (1986) Characteristics of maximum-value composite images from temporal AVHRR data. Int J Remote Sens 7(11):1417–1434

Holling CS (1973) Resilience and stability of ecological systems. Annu Rev Ecol Syst 4:1–23

Horion S, Fensholt R, Tagesson T, Ehammer A (2014) Using earth observation-based dry season NDVI trends for assessment of changes in tree cover in the Sahel. Int J Remote Sens 35(7):2493–2515

Hubert B, Rosegrant M, van Boekel M, Ortiz R (2010) The future of food: scenarios for 2050. Crop Sci 50(Suppl 1):33–50

Huete A, Didan K, Miura T, Rodriguez EP, Gao X, Ferreira LG (2002) Overview of the radiometric and biophysical performance of the MODIS vegetation indices. Remote Sens Environ 83(1):195–213

Huffman GJ, Adler RF, Bolvin DT, Gu G (2009) Improving the global precipitation record: GPCP version 2.1. Geophys Res Lett 36(17):L17808

Hutchinson C (1991) Uses of satellite data for famine early warning in sub-Saharan Africa. Int J Remote Sens 12(6):1405–1421

Isaev A, Korovin G, Bartalev S, Ershov D, Janetos A, Kasischke E, Shugart H, French N, Orlick B, Murphy T (2002) Using remote sensing to assess Russian forest fire carbon emissions. Clim Change 55(1–2):235–249

ISRIC (2008a) Enhanced soil information for LADA partner countries, scale 1:1 million. World Soil Information (ISRIC), Wageningen

ISRIC (2008b) Global SOTER landform classification scale 1:1 million. World Soil Information (ISRIC), Wageningen

Ito A, Oikawa T (2002) A simulation model of the carbon cycle in land ecosystems (Sim-CYCLE): a description based on dry-matter production theory and plot-scale validation. Ecol Model 151(2):143–176

Ivits E, Cherlet M, Sommer S, Mehl W (2012a) Ecosystem Functional Units characterized by satellite observed phenology and productivity gradients: a case study for Europe. Ecol Indic 27:17–28

Ivits E, Cherlet M, Toth G, Sommer S, Mehl W, Vogt J, Micale F (2012b) Combining satellite derived phenology with climate data for climate change impact assessment. Global Planet Change 88–89:85–97

Jackson TJ, Chen D, Cosh M, Li F, Anderson M, Walthall C, Doriaswamy P, Hunt E (2004) Vegetation water content mapping using Landsat data derived normalized difference water index for corn and soybeans. Remote Sens Environ 92(4):475–482

Jacquin A, Sheeren D, Lacombe J-P (2010) Vegetation cover degradation assessment in Madagascar savanna based on trend analysis of MODIS NDVI time series. Int J Appl Earth Obs Geoinf 12:S3–S10

James M, Kalluri SN (1994) The Pathfinder AVHRR land data set: an improved coarse resolution data set for terrestrial monitoring. Int J Remote Sens 15(17):3347–3363

Jensen J (2007) Remote sensing of the environment. Pearson Prentice Hall, Upper Saddle River

Jiang Z, Huete AR, Didan K, Miura T (2008) Development of a two-band enhanced vegetation index without a blue band. Remote Sens Environ 112(10):3833–3845

Joiner J, Yoshida Y, Vasilkov A, Middleton E (2011) First observations of global and seasonal terrestrial chlorophyll fluorescence from space. Biogeosciences 8(3):637–651

Joiner J, Yoshida Y, Vasilkov A, Middleton E, Campbell P, Kuze A (2012) Filling-in of near-infrared solar lines by terrestrial fluorescence and other geophysical effects: simulations and space-based observations from SCIAMACHY and GOSAT. Atmos Meas Tech 5(4):809–829

Joiner J, Guanter L, Lindstrot R, Voigt M, Vasilkov A, Middleton E, Huemmrich K, Yoshida Y, Frankenberg C (2013) Global monitoring of terrestrial chlorophyll fluorescence from moderate spectral resolution near-infrared satellite measurements: methodology, simulations, and application to GOME-2. Atmos Meas Tech Discuss 6(2):3883–3930

Jones P, Harris I (2013) CRU TS3.21: Climatic Research Unit (CRU) Time-Series (TS) Version 3.21 of high resolution gridded data of month-by-month variation in climate (Jan 1901–Dec 2012). University of East Anglia Climatic Research Unit (CRU), NCAS British Atmospheric Data Centre. http://badc.nerc.ac.uk/view/badc.nerc.ac.uk

Jordan CF (1969) Derivation of leaf-area index from quality of light on the forest floor. Ecology 50:663–666

Jong R et al. (2012) Trend changes in global greening and browning: contribution of short-term trends to longterm change. Glob Chang Biol 18(2):642–655

Justice CO, Townshend J, Holben B, Tucker CJ (1985) Analysis of the phenology of global vegetation using meteorological satellite data. Int J Remote Sens 6(8):1271–1318

Karnieli A, Dall'Olmo G (2003) Remote-sensing monitoring of desertification, phenology, and droughts. Manag Environ Qual 14(1):22–38

Karnieli A, Agam N, Pinker RT, Anderson M, Imhoff ML, Gutman GG, Panov N, Goldberg A (2010) Use of NDVI and land surface temperature for drought assessment: merits and limitations. J Climate 23(3):618–633

Khorram S, Koch FH, van der Wiele CF, Nelson SA (2012) Remote sensing. Springer, New York

Kinzig AP, Ryan P, Etienne M, Allison H, Elmqvist T, Walker BH (2006) Resilience and regime shifts: assessing cascading effects. Ecol Soc 11(1):23

Krupenikov IA, Dent D, Boincean BP (2011) The black earth: ecological principles for sustainable agriculture on chernozem soils, vol 10. Springer, Dordrecht

Lambin EF, Ehrlich D (1997) Land-cover changes in sub-Saharan Africa (1982–1991): application of a change index based on remotely sensed surface temperature and vegetation indices at a continental scale. Remote Sens Environ 61(2):181–200

Lambin EF, Geist HJ, Lepers E (2003) Dynamics of land-use and land-cover change in tropical regions. Annu Rev Environ Resour 28(1):205–241

Landmann T, Dubovyk O (2014) Spatial analysis of human-induced vegetation productivity decline over eastern Africa using a decade (2001–2011) of medium resolution MODIS time-series data. Int J Appl Earth Obs Geoinf 33:76–82

Lanorte, A, et al. (2014) Fisher–Shannon information plane analysis of SPOT/VEGETATION Normalized Difference Vegetation Index (NDVI) time series to characterize vegetation recovery after fire disturbance. International Journal of Applied Earth Observation and Geoinformation, 26, 441–446

Le Houerou HN (1984) Rain use efficiency: a unifying concept in arid-land ecology. J Arid Environ 7(3):213–247

Le QB, Tamene L, Vlek PLG (2012) Multi-pronged assessment of land degradation in West Africa to assess the importance of atmospheric fertilization in masking the processes involved. Global Planet Change 92–93:71–81. doi:10.1016/j.gloplacha.2012.05.003

Le QB, Nkonya E, Mirzabaev A (2014) Biomass productivity-based mapping of global land degradation hotspots. ZEF-Discussion Papers on Development Policy (193)

Leon JRR, van Leeuwen WJ, Casady GM (2012) Using MODIS-NDVI for the modeling of post-wildfire vegetation response as a function of environmental conditions and pre-fire restoration treatments. Remote Sens 4(3):598–621

Li J, Lewis J, Rowland J, Tappan G, Tieszen L (2004) Evaluation of land performance in Senegal using multi-temporal NDVI and rainfall series. J Arid Environ 59(3):463–480

Liang S (2005) Quantitative remote sensing of land surfaces, vol 30. Wiley, New York

Liu W, Juárez RN (2001) ENSO drought onset prediction in northeast Brazil using NDVI. Int J Remote Sens 22(17):3483–3501

Lobell D, Lesch S, Corwin D, Ulmer M, Anderson K, Potts D, Doolittle J, Matos M, Baltes M (2010) Regional-scale assessment of soil salinity in the Red River Valley using multi-year MODIS EVI and NDVI. J Environ Qual 39(1):35–41

Lunetta RS, Knight JF, Ediriwickrema J, Lyon JG, Worthy LD (2006) Land-cover change detection using multi-temporal MODIS NDVI data. Remote Sens Environ 105(2): 142–154

Malak DA, Pausas JG (2006) Fire regime and post-fire Normalized Difference Vegetation Index changes in the eastern Iberian peninsula (Mediterranean basin). Int J Wildland Fire 15(3):407–413

Mancino G, Nolè A, Ripullone F, Ferrara A (2014) Landsat TM imagery and NDVI differencing to detect vegetation change: assessing natural forest expansion in Basilicata, southern Italy. iForest-Biogeosci Forestry 7(2):76

Mas J-F (1999) Monitoring land-cover changes: a comparison of change detection techniques. Int J Remote Sens 20(1):139–152

Mather P, Koch M (2011) Computer processing of remotely-sensed images: an introduction. Wiley, Chichester

Mayaux P, Eva H, Brink A, Achard F, Belward A (2008) Remote sensing of land-cover and land-use dynamics. In: Chuvieco E (ed) Earth observation of global change. Springer, New York, pp 85–108

Mbow C, Fensholt R, Nielsen TT, Rasmussen K (2014) Advances in monitoring vegetation and land use dynamics in the Sahel. Geografisk Tidsskrift-Danish J Geogr 114(1):84–91

MEA (2005) Millennium ecosystem assessment: ecosystems and human well-being. Statement from the Board. Millennium ecosystem assessment. Full Report. http://www.millenniumassessment.org

Metternicht G, Zinck J (2003) Remote sensing of soil salinity: potentials and constraints. Remote Sens Environ 85(1):1–20

Mishra AK, Singh VP (2010) A review of drought concepts. J Hydrol 391(1–2):202–216. doi:http://dx.doi.org/10.1016/j.jhydrol.2010.07.012

Mishra U, Lal R, Liu D, Van Meirvenne M (2010) Predicting the spatial variation of the soil organic carbon pool at a regional scale. Soil Sci Soc Am J 74(3):906–914

Mitchell TD, Jones PD (2005) An improved method of constructing a database of monthly climate observations and associated high-resolution grids. Int J Climatol 25(6):693–712

Mitchell JE, Roundtable SR (2010) Criteria and indicators of sustainable rangeland management. University of Wyoming/Cooperative Extension Service, Laramie

Monteith J, Moss C (1977) Climate and the efficiency of crop production in Britain [and discussion]. Philos Trans Royal Soc London B, Biol Sci 281(980):277–294

Morton DC, Nagol J, Carabajal CC, Rosette J, Palace M, Cook BD, Vermote EF, Harding DJ, North PR (2014) Amazon forests maintain consistent canopy structure and greenness during the dry season. Nature 506(7487):221–224

Mulianga B, Bégué A, Simoes M, Clouvel P, Todoroff P (2013) Estimating potential soil erosion for environmental services in a sugarcane growing area using multisource remote sensing data. In: SPIE remote sensing. International Society for Optics and Photonics, pp 88871W-88810

Myneni RB, Hall FG, Sellers PJ, Marshak AL (1995) The interpretation of spectral vegetation indexes. IEEE Trans Geosci Remote Sens 33(2):481–486

Myneni R. B. et al. (2014) Attribution of global vegetation photosynthetic capacity from 1982 to 2014. Global Change Biology (in review)

Nemani RR, Keeling CD, Hashimoto H, Jolly WM, Piper SC, Tucker CJ, Myneni RB, Running SW (2003) Climate-driven increases in global terrestrial net primary production from 1982 to 1999. Science 300(5625):1560–1563

Nkonya E, Gerber N, Baumgartner P, Von Braun J, De Pinto A, Graw V, Kato E, Kloos J, Walter T (2011) The economics of desertification, land degradation, and drought: toward an integrated global assessment. ZEF Discussion Papers on Development Policy

Novella NS, Thiaw WM (2013) African rainfall climatology version 2 for famine early warning systems. J Appl Meteorol Climatol 52(3):588–606

Oindo BO, de By RA, Skidmore AK (2000) Interannual variability of NDVI and bird species diversity in Kenya. Int J Appl Earth Obs Geoinf 2(3):172–180

Oldeman L, Hakkeling R, Sombroek W (1990) World map on status of human-induced soil degradation (GLASOD) UNEP. ISRIC, Nairobi

Olsson L, Eklundh L, Ardö J (2005) A recent greening of the Sahel—trends, patterns and potential causes. J Arid Environ 63(3):556–566

Orr BJ (2011) Scientific review of the UNCCD provisionally accepted set of impact indicators to measure the implementation of strategic objectives 1, 2 and 3. Office of Arid Lands Studies. University of Arizona, Arizona

Orr BJ (2014) Personal communication on: the use of normalized difference vegetation index for land degradation assessment—country applications. Interviewed on 13 Oct 2014

Pan C, Zhao H, Zhao X, Han H, Wang Y, Li J (2013) Biophysical properties as determinants for soil organic carbon and total nitrogen in grassland salinization. PLoS One 8(1), e54827

Pearson RL, Miller LD (1972) Remote mapping of standing crop biomass for estimation of the productivity of the shortgrass prairie. Remote Sens Environ VIII:1355

Pedelty J, Devadiga S, Masuoka E, Brown M, Pinzon J, Tucker C, Roy D, Ju J, Vermote E, Prince S (2007) Generating a long-term land data record from the AVHRR and MODIS instruments. In: Proceedings of the IEEE 2007 International Geoscience and Remote Sensing Symposium, Barcelona, Spain, pp 1021–1025

Perrings C (1998) Resilience in the dynamics of economy-environment systems. Environ Resour Econ 11(3–4):503–520

Pettorelli N (2013) The normalized difference vegetation index. Oxford University Press, Oxford

Pettorelli N, Vik JO, Mysterud A, Gaillard J-M, Tucker CJ, Stenseth NC (2005) Using the satellite-derived NDVI to assess ecological responses to environmental change. Trends Ecol Evol 20(9):503–510

Pettorelli N, Safi K, Turner W (2014) Satellite remote sensing, biodiversity research and conservation of the future. Philos Trans R Soc, B 369(1643):20130190

Piao S, Fang J, Ciais P, Peylin P, Huang Y, Sitch S, Wang T (2009) The carbon balance of terrestrial ecosystems in China. Nature 458(7241):1009–1013

Pinzon J, Tucker C (2014) A non-stationary 1981–2012 AVHRR NDVI3G time series. Remote Sens 6(8):6929–6960

Platonov A, Noble A, Kuziev R (2013) Soil salinity mapping using multi-temporal satellite images in agricultural fields of syrdarya province of Uzbekistan. In: Shahid SA, Abdelfattah MA, Taha FK (eds) Developments in soil salinity assessment and reclamation: innovative thinking and use of marginal soil and water resources in irrigated agriculture. Springer, London, pp 87–98

Porter JR, Xie L, Challinor A, Cochrane K, Mark Howden M, Iqbal MM, Lobell D, Travasso MT (2014) Food security and food production systems. In: Field CB, Barros VR, Dokken DJ et al (eds) Climate change 2014: impacts, adaptation, and vulnerability. Part A: global and sectoral aspects. Contribution of Working Group II to the Fifth Assessment Report of the Intergovernmental Panel on Climate Change, vol 1. Cambridge University Press, Cambridge

Prasannakumar V, Vijith H, Abinod S, Geetha N (2012) Estimation of soil erosion risk within a small mountainous sub-watershed in Kerala, India, using Revised Universal Soil Loss Equation (RUSLE) and geo-information technology. Geosci Front 3(2):209–215

Prince SD, Goward SN (1995) Global primary production: a remote sensing approach. J Biogeogr 22:815–835

Prince SD, Colstoun D, Brown E, Kravitz L (1998) Evidence from rain-use efficiencies does not indicate extensive Sahelian desertification. Glob Chang Biol 4(4):359–374

Prince S, Becker-Reshef I, Rishmawi K (2009) Detection and mapping of long-term land degradation using local net production scaling: application to Zimbabwe. Remote Sens Environ 113(5):1046–1057

Propastin P, Kappas M, Muratova N (2008) Application of geographically weighted regression analysis to assess human-induced land degradation in a dry region of Kazakhstan. In: shaping the change, XXIII FIG congress

Purkis SJ, Klemas VV (2011) Remote sensing and global environmental change. Wiley, Oxford

Quarmby N, Milnes M, Hindle T, Silleos N (1993) The use of multi-temporal NDVI measurements from AVHRR data for crop yield estimation and prediction. Int J Remote Sens 14(2):199–210

Renschler CS, Frazier A, Arendt L, Cimellaro G-P, Reinhorn AM, Bruneau M (2010) A framework for defining and measuring resilience at the community scale: the PEOPLES resilience framework. MCEER, Buffalo

Rienecker MM, Suarez MJ, Gelaro R, Todling R, Bacmeister J, Liu E, Bosilovich MG, Schubert SD, Takacs L, Kim G-K (2011) MERRA: NASA's modern-era retrospective analysis for research and applications. J Climate 24(14):3624–3648

Rockström J, Steffen W, Noone K, Persson Å, Chapin FS, Lambin EF, Lenton TM, Scheffer M, Folke C, Schellnhuber HJ (2009) A safe operating space for humanity. Nature 461(7263):472–475

Ross J (1981) The radiation regime and architecture of plant stands, vol 3. Springer, Dordrecht

Rouse J Jr, Haas R, Schell J, Deering D (1974) Monitoring vegetation systems in the Great Plains with ERTS. NASA Spec Publ 351:309

Rudolf B, Beck C, Grieser J, Schneider U (2005) Global precipitation analysis products of the GPCC. Climate monitoring—Tornadoklimatologie–Aktuelle Ergebnisse des Klimamonitorings, Germany, pp 163–170

Running SW, Nemani RR, Heinsch FA, Zhao M, Reeves M, Hashimoto H (2004) A continuous satellite-derived measure of global terrestrial primary production. Bioscience 54(6):547–560

Saatchi SS, Harris NL, Brown S, Lefsky M, Mitchard ET, Salas W, Zutta BR, Buermann W, Lewis SL, Hagen S (2011) Benchmark map of forest carbon stocks in tropical regions across three continents. Proc Natl Acad Sci 108(24):9899–9904

Safriel UN (2007) The assessment of global trends in land degradation. In: Sivakumar MK, Ndiang'ui N (eds) Climate and land degradation, Environmental science and engineering: environmental science. Springer, Berlin, pp 1–38

Saleska SR, Didan K, Huete AR, Da Rocha HR (2007) Amazon forests green-up during 2005 drought. Science 318(5850):612

Scheftic W, Zeng X, Broxton P, Brunke M (2014) Intercomparison of seven NDVI products over the United States and Mexico. Remote Sens 6(2):1057–1084

Schlerf M, Atzberger C, Hill J (2005) Remote sensing of forest biophysical variables using HyMap imaging spectrometer data. Remote Sens Environ 95(2):177–194

Schneider U et al. (2008) Global precipitation analysis products of the GPCC. Global Precipitation Climatology Centre (GPCC), DWD, Internet Publikation, 112

Sellers P, Tucker C, Collatz G, Los S, Justice C, Dazlich D, Randall D (1994) A global 1 by 1 NDVI data set for climate studies. Part 2: the generation of global fields of terrestrial biophysical parameters from the NDVI. Int J Remote Sens 15(17):3519–3545

Shalaby A, Tateishi R (2007) Remote sensing and GIS for mapping and monitoring land cover and land-use changes in the Northwestern coastal zone of Egypt. Appl Geogr 27(1):28–41

Sietse O (2010) ISLSCP II FASIR-adjusted NDVI, 1982–1998. ISLSCP Initiative II Collection Data set Available on-line [http://daac.ornl.gov/] from Oak Ridge National Laboratory Distributed Active Archive Center, Oak Ridge, TN, USA

Simoniello T, Lanfredi M, Liberti M, Coppola R, Macchiato M (2008) Estimation of vegetation cover resilience from satellite time series. Hydrol Earth Syst Sci Discuss 5(1):511–546

Sivakumar MV, Stefanski R (2007) Climate and land degradation—an overview. In: Climate and land degradation. Springer, New York, pp 105–135

Sobrino J, Raissouni N (2000) Toward remote sensing methods for land cover dynamic monitoring: application to Morocco. Int J Remote Sens 21(2):353–366

Sommer S, Zucca A, Grainger M, Cherlet R, Zougmore Y, Sokona J, Hill R, Della-Peruta J, Wang G (2011) Application of indicator systems for monitoring and assessment of desertification from national to global scales. Land Degrad Dev 22:184–197

Sonneveld B, Dent D (2009) How good is GLASOD? J Environ Manage 90(1):274–283

Steiniger S, Hunter AJS (2013) The 2012 free and open source GIS software map—a guide to facilitate research, development, and adoption. Comput Environ Urban Syst 39:136–150. doi:http://dx.doi.org/10.1016/j.compenvurbsys.2012.10.003

Sternberg T, Tsolmon R, Middleton N, Thomas D (2011) Tracking desertification on the Mongolian steppe through NDVI and field-survey data. Int J Digital Earth 4(1):50–64

Stocking M, Murnaghan N (2000) Land degradation–guidelines for field assessment. Overseas Development Group, University of East Anglia, Norwich, p 121

Stow DA, Hope A, McGuire D, Verbyla D, Gamon J, Huemmrich F, Houston S, Racine C, Sturm M, Tape K (2004) Remote sensing of vegetation and land-cover change in Arctic Tundra Ecosystems. Remote Sens Environ 89(3):281–308

Strand H, Höft R, Strittholt J, Horning N, Miles L, Fosnight E, Turner W (2007) Sourcebook on remote sensing and biodiversity indicators, vol Technical Series no. 32. Secretariat of the Convention on Biological Diversity, Montreal

Strittholt J, Steininger M (2007) Trends in selected biomes, habitats, and ecosystems: forests. In: Strand H, Höft R, Strittholt J et al (eds) Sourcebook on remote sensing and biodiversity indicators, vol Technical Series No. 32. Secretariat of the Convention on Biological Diversity, Montreal, p 203

Strzepek K, Boehlert B (2010) Competition for water for the food system. Philos Trans R Soc, B 365(1554):2927–2940

Symeonakis E, Drake N (2004) Monitoring desertification and land degradation over sub-Saharan Africa. Int J Remote Sens 25(3):573–592

Thorén H, Persson J (2014) A concept of resilience for sustainability science: some remarks on ostensive and stipulative definitions. Soc Nat Resour 11(1):64–74

Tilman D (2010) Understanding the present and projecting the future of global food demand. Paper presented at the AAAS Annual Meeting 2010, San Diego

Tilman D, Cassman K, Matson P, Naylor R, Polasky S (2002) Agricultural sustainability and intensive production practices. Nature 418(6898):671–677

Townshend JR, Goff TE, Tucker CJ (1985) Multitemporal dimensionality of images of normalized difference vegetation index at continental scales. IEEE Trans Geosci Remote Sens 6:888–895

Townshend JR, Masek JG, Huang C, Vermote EF, Gao F, Channan S, Sexton JO, Feng M, Narasimhan R, Kim D (2012) Global characterization and monitoring of forest cover using Landsat data: opportunities and challenges. Int J Digital Earth 5(5):373–397

Tucker CJ (1979) Red and photographic infrared linear combinations for monitoring vegetation. Remote Sens Environ 8(2):127–150

Tucker CJ, Garratt MW (1977) Leaf optical system modeled as a stochastic process. Appl Optics 16(3):635–642

Tucker CJ, Nicholson SE (1999) Variations in the size of the Sahara Desert from 1980 to 1997. Ambio 28:587–591

Tucker CJ, Holben BN, Elgin JH Jr, McMurtrey JE III (1981) Remote sensing of total dry-matter accumulation in winter wheat. Remote Sens Environ 11:171–189

Tucker C, Vanpraet CL, Sharman M, Van Ittersum G (1985) Satellite remote sensing of total herbaceous biomass production in the Senegalese Sahel: 1980–1984. Remote Sens Environ 17(3):233–249

Tucker CJ, Pinzon JE, Brown ME, Slayback DA, Pak EW, Mahoney R, Vermote EF, El Saleous N (2005) An extended AVHRR 8-km NDVI dataset compatible with MODIS and SPOT vegetation NDVI data. Int J Remote Sens 26(20):4485–4498

Turner BL, Meyer WB (1994) Global land-use and land-cover change: an overview. In: Changes in land use and land cover: a global perspective, vol 4. Cambridge University Press, Cambridge, p 3

Turner W, Spector S, Gardiner N, Fladeland M, Sterling E, Steininger M (2003) Remote sensing for biodiversity science and conservation. Trends Ecol Evol 18(6):306–314

Udelhoven T, Stellmes M (2007) Changes in land surface conditions on the Iberian Peninsula (1989 to 2004) detected by means of time series analysis from hypertemporal remote sensing data, Proceedings of MultiTemp 2007 International Workshop on the Analysis of Multi-Temporal Remote Sensing Images. MultiTemp, Leuven, Belgium, pp 1–6

UN (2013) World population prospects: the 2012 revision, highlights and advance tables. Working Paper, vol ESA/P/WP.228. United Nations, Department of Economic and Social Affairs, Population Division, New York

UNCCD (1994) Elaboration of an international convention to combat desertification in countries experiencing serious drought and/or desertification, particularly in Africa. United Nations Convention to Combat Desertification, Paris

UNEP (2007) Global environment outlook GEO4—environment for development. United Nations Environmental Programme (UNEP), Nairobi

UNEP (2012a) Environment for the future we want. In: GEO 5—Global Environment Outlook. United Nations Environment Programme, Nairobi

UNEP (2012b) Sahel atlas of changing landscapes: tracing trends and variations in vegetation cover and soil condition. United Nations Environment Programme, Nairobi

United Nations (2011) Millennium development goals report 2011. ISBN 978-92-1-101244-6. http://www.refworld.org/docid/4e42118b2.html. Accessed 1 Oct 2014

Vacik H, Wolfslehner B, Seidl R, Lexer MJ (2007) Integrating the DPSIR approach and the analytic network process for the assessment of forest management strategies. In: Reynolds K, Thomson A, Köhl M, Shannon M, Ray D, Rennolls K (eds) Sustainable forestry: from monitoring and modelling to knowledge management and policy science. CAB International, Cambridge, pp 393–411

Veldkamp A, Lambin EF (2001) Predicting land-use change. Agr Ecosyst Environ 85(1):1–6

Verbesselt J, Hyndman R, Newnham G, Culvenor D (2010a) Detecting trend and seasonal changes in satellite image time series. Remote Sens Environ 114(1):106–115

Verbesselt J, Hyndman R, Zeileis A, Culvenor D (2010b) Phenological change detection while accounting for abrupt and gradual trends in satellite image time series. Remote Sens Environ 114(12):2970–2980

Verón SR, Oesterheld M, Paruelo JM (2005) Production as a function of resource availability: slopes and efficiencies are different. J Veg Sci 16(3):351–354

Verrelst J, Koetz B, Kneubühler M, Schaepman M (2006) Directional sensitivity analysis of vegetation indices from multi-angular Chris/PROBA data. In: ISPRS Commission VII Mid-term symposium, pp 677–683

Vlek P, Le Q, Tamene L (2010) Assessment of land degradation, its possible causes and threat to food security in Sub-Saharan Africa. In: Lal R, Stewart BA (eds) Food security and soil quality, Advances in soil science. Taylor & Francis, Boca Raton, pp 57–86

Vogt JV, Safriel U, Von Maltitz G, Sokona Y, Zougmore R, Bastin G, Hill J (2011) Monitoring and assessment of land degradation and desertification: towards new conceptual and integrated approaches. Land Degrad Dev 22(2):150–165. doi:10.1002/ldr.1075

Walker B, Holling CS, Carpenter SR, Kinzig A (2004) Resilience, adaptability and transformability in social—ecological systems. Ecol Soc 9(2):5

Walker B, Abel N, Andreoni F, Cape J, Murdock H, Norman C (2014) General resilience: a discussion paper based on insights from a catchment management area workshop in South Eastern Australia. Resilience Alliance, Australia

Wang J, Sammis TW, Gutschick VP, Gebremichael M, Dennis SO, Harrison RE (2010) Review of satellite remote sensing use in forest health studies. Open Geogr J 3:28–42

Warren A (2002) Land degradation is contextual. Land Degrad Dev 13(6):449–459

Wessels K (2009) Letter to the editor: comments on 'Proxy global assessment of land degradation' by Bai et al. (2008). Soil Use Manag 25:91–92

Wessels KJ, Prince SD, Frost PE, van Zyl D (2004) Assessing the effects of human-induced land degradation in the former homelands of northern South Africa with a 1 km AVHRR NDVI time-series. Remote Sens Environ 91(1):47–67. doi:http://dx.doi.org/10.1016/j.rse.2004.02.005

Wessels KJ, Prince SD, Malherbe J, Small J, Frost PE, VanZyl D (2007) Can human-induced land degradation be distinguished from the effects of rainfall variability? A case study in South Africa. J Arid Environ 68(2):271–297

Wessels K, Van Den Bergh F, Scholes R (2012) Limits to detectability of land degradation by trend analysis of vegetation index data. Remote Sens Environ 125:10–22

Xie Y, Sha Z, Yu M (2008) Remote sensing imagery in vegetation mapping: a review. J Plant Ecol 1(1):9–23

Xu D, Kang X, Qiu D, Zhuang D, Pan J (2009) Quantitative assessment of desertification using landsat data on a regional scale–a case study in the Ordos plateau, China. Sensors 9(3):1738–1753

Yan H, Wang S, Wang C, Zhang G, Patel N (2005) Losses of soil organic carbon under wind erosion in China. Glob Chang Biol 11(5):828–840

Yeqiao W (2011) Remote sensing of protected lands. Remote sensing applications series. CRC Press, Boca Raton, pp 1–26. doi: 10.1201/b11453-210.1201/b11453-2

Yin H, Udelhoven T, Fensholt R, Pflugmacher D, Hostert P (2012) How normalized difference vegetation index (ndvi) trends from advanced very high resolution radiometer (AVHRR) and système probatoire d'observation de la terre vegetation (spot vgt) time series differ in agricultural areas: an inner mongolian case study. Remote Sens 4(11):3364–3389

Yoshida Y, Joiner J, Tucker C, Berry J, Lee J-E, Walker G, Reichle R, Koster R, Lyapustin A, Wang Y (2014) The 2010 Russian drought impact on satellite measurements of solar-induced chlorophyll fluorescence: insights from modeling and comparisons with the Normalized Differential Vegetation Index (NDVI). Remote Sens Environ 166:163–177

Yuan D, Elvidge C (1998) NALC land cover change detection pilot study: Washington DC area experiments. Remote Sens Environ 66(2):166–178

Zargar A, Sadiq R, Naser B, Khan FI (2011) A review of drought indices. Environ Rev 19(NA):333–349. doi:10.1139/a11-013

Zhang Y, Peng C, Li W, Fang X, Zhang T, Zhu Q, Chen H, Zhao P (2013) Monitoring and estimating drought-induced impacts on forest structure, growth, function, and ecosystem services using remote-sensing data: recent progress and future challenges. Environ Rev 21(2):103–115. doi:10.1139/er-2013-0006

Zhou P, Luukkanen O, Tokola T, Nieminen J (2008) Effect of vegetation cover on soil erosion in a mountainous watershed. Catena 75(3):319–325

Zinck A, Metternich G (2008) Soil salinity and salinization hazard. In: Metternich G, Zinck A (eds) Remote sensing of soil salinization: impact on land management. CRC Press, Boca Raton

Printed in the United States
By Bookmasters